viation Safety and Automation Implications.

Richard H. Green

ABSTRACT

Since the first aircraft accident was attributed to the improper use of automation

technology in 1996, the aviation community has recognized that the benefits of flight

deck technology also have negative unintended consequences from both the technology

itself and the human interaction required to implement and operate it. This mixed

methods study looks at the relationship of technology to the severity of aircraft mishaps

and the policy implications resulting from those relationships in order to improve safety

of passenger carrying aircraft in the United States National Airspace System. U.S. mishap

data from the National Transportation Safety Board and the Aviation Safety Reporting

System was collected covering aircraft mishaps spanning the last twenty years. An

ordinal regression was used to determine which types of flight deck technology played a

significant role in the severity of aircraft mishaps ranging from minor to catastrophic.

Using this information as a focal point, a qualitative analysis was undertaken to analyze

the mechanisms for that impact, the effect of existing policy guidance relating to the use

of technology, and the common behaviors not addressed by policy that provide a venue to

address aviation safety. Some areas of current policy were found to be effective, while

multiple areas of opportunity for intervention were uncovered at the various levels of

aircraft control including the organizational, the supervisory, the preparatory, and the execution level that suggest policy adjustments that may be made to reduce incidence of

control failure caused by cockpit automation.

TABLE OF CONTENTS

CHAPTER

LIST OF TABLES

TABLE

LIST OF FIGURES

FIGURE

CHAPTER I

INTRODUCTION

This research study evaluates the role of policy regarding the use of automation in flight on the safety of aviation in the United States. Specifically, following the investigation of several high-profile aircraft mishaps that identified misuse or overuse of automation at inappropriate times during flight, several best practices have developed among airline operators as to the acceptable use of flight automation. However, there has never been a coordinated governmental policy on the use of automated equipment, procedures, or communication promulgated by the Federal Aviation Administration (FAA), the International Civil Aviation Organization (ICAO), or industry workgroups.

While there has been considerable research among social scientists on the impact of human error and misunderstanding on automated systems, and while great progress has been made in the area of human factor consideration in designing airplanes and the systems used to control them, there has been little research on the public policy aspect of these considerations. Even though the FAA has established a research group specifically dedicated to the impact of human factor considerations on flying (Chandra & Grayhem, 2012), there has been no subsequent discussion on the policy changes those results should dictate.

Because of the relative infrequency of major aircraft mishaps, coupled with the often high toll in injury and death, commercial aircraft accidents garner a high level of interest from both the public and from governmental authorities. Analyzing the safety of any system related to this area of transportation requires both a thorough analysis of the human component of the process as well as an evaluation of the appropriate level of control that should be ceded to an automated system. Such a system is subject to all the failings that may be associated with automation that possesses a limited strategic view, a lack of human capacity to think, and that will face unforeseen situations that may occur beyond the scope of the original programming design (Berry & Sawyer, 2013; Inagaki, 2003). Currently there are numerous sources of guidance from the FAA in the forms of the Code of Federal Regulations (CFR), Federal Aviation Regulations (FARs), Advisory Circulars (ACs), and orders but these are limited to very specific instructions on qualifications for pilots in specific types of activities and to certification requirements for the equipment being used, not to policy guidance on the appropriateness of use (Chandra, Grayhem, & Butchibabu, 2012). Given that the goal of automation is to improve pilot situational awareness, efficiency, and flying safety (Spirkovska, 2004), and given the ability of new technology to change the environment itself and to introduce breakdowns in control (Dalcher, 2007), an investigation into the appropriate role of automation policy in aviation is needed.

Background

On the evening of Dec 20, 1995, American Airlines flight 965 (AA965) was flying its regularly scheduled route from Miami, Florida, to Cali, Colombia with 163 passengers and crew on board. It was a dark and nearly moonless night, but otherwise skies were clear with only a few scattered clouds. The crew was running behind schedule and was offered an opportunity to shave a few minutes off their arrival time by accepting a direct clearance to the ROZO nondirectional beacon (NDB) navigational aid. The experienced captain programmed the identifier code for ROZO—"R"—into the flight management computer (FMC). Unfortunately, the letter "R" was also the identifier for several other navigational aids in the computer database and the first option offered and selected was actually a geographical point named ROMEO, 132 nautical miles northeast of their location. Before the crew could catch this error, the plane struck near the summit of El Deluvio at the 8,900 elevation level. Only four passengers survived the accident (Dalcher, 2007; Simmon, 1998).

This aircraft mishap represents one of the first times that investigators identified the use—or overuse—of cockpit automation as a significant causal factor in an accident (Lintern, 2000). Since then there have been multiple others, with the most commonly referenced highlighted in Appendix A. Specifically the final accident investigation report for AA965 identified several specific actions as translated by the Flight Safety Foundation:

- The lack of situational awareness of the flight crew regarding vertical navigation, proximity to terrain and the relative location of critical radio aids.

- Failure of the flight crew to revert to basic radio navigation at the time when the FMS [Flight Management System]-assisted navigation became confusing and demanded an excessive workload in a critical phase of the flight.

- FMS logic that dropped all intermediate fixes from the display(s) in the event of execution of a direct routing.

- FMS-generated navigational information that used a different naming convention from that published in navigational charts. (Simmon, 1998, p. 1)

In 1903, the Wright brothers finally achieved the long held human dream of powered flight. In more than a century since then, air travel has become an invaluable part of modern life. By 2004, U.S. air carriers were logging more than eight billion miles of flight (Hendrickson, 2009). The prevalence of people traveling by aircraft transportation shows no signs of slowing down. By 2012, 2.9 billion passengers boarded an airplane for either business or leisure world-wide (Marzuoli et al., 2014), and the ICAO, the arm of the United Nations tasked with worldwide regulation of aviation, forecasts an annual growth in world travel of 5% through at least 2020 (Hollnagel, 2007). With more and more aircraft competing for limited space at airports and in the airspace designed to funnel airplanes into those airports, the need for efficiency in controlling and routing those airplanes is evident. This need coupled with rapid advances in technology has led to a significant increase in cockpit automation levels even since the events of Cali in 1995.

In the first aircraft, the Wright brothers gave the individual full control and those early pilots were perceived as daredevils. This role is far removed from the perception most passengers have of commercial pilots today as professionals managing highly complex, computerized systems. The transformation in the role of the pilot started almost immediately. The Wrights recognized their invention as inherently unstable and in 1905 Orville started work on a stability augmentation device. The device was ready by 1913 (Koeppen, 2012). By the next year, Lawrence Sperry developed a two-gyroscope system that would sense and make corrections to deviations from normal flight. This first autopilot system was the direct ancestor of the highly complex inertial navigation system used in the Apollo space missions and in today's modern aircraft (Koeppen, 2012).

As with other parts of our daily life, automation represents one of the major trends in aviation today. When aircrews first started flying in instrument-required conditions (clouds, nighttime, outside of visual range of landmarks, etc.) they used printed charts that provided them the primary source of information needed to fly the procedures that would allow safe landing and crosschecked that data by manual reference to their flight instrumentation (Chandra et al., 2012). Over time, this method proved inadequate for both safety of flight and efficiency of traffic flow. By the end of the 1960s, the economic possibilities of expanded passenger traffic placed a high priority on increasing capacity through the use of increased automation, as aircraft sought to increase aircraft utilization by shortening time enroute, positioning to land, and time on the ground between flights (Amalberti, 1999; Marzuoli et al., 2014). The military application of electronic flight displays was then adapted for commercial transport category airplanes, including the use of cathode ray tubes for display (Koeppen, 2012). By the 1970s, the average transport aircraft had more than 100 cockpit instruments and controls, all of which competed for pilot attention and real estate in the cockpit display (Koeppen, 2012). Economic demands, coupled with advances in technology that include inertial guidance hardware, global positioning systems, computerized instrumentation, and digital satellite communications, have significantly changed the role of the pilot in flying and navigating the aircraft. Most recently, the incorporation of iPads® into the cockpit has competed for the attention of pilots while flying, navigating, and communicating (Joslin, 2013).

Automation has steadily advanced as technology has allowed for computerized control of many formerly manually performed physical, perceptual, and even cognitive tasks (Endsley, 1996), while technological advances allow for more and more automated functions on the flight deck under dynamic situations with the potential for significant consequences (Sarter & Woods, 1994).

The newest generation of computerized flight decks has been christened with the nickname of *glass cockpits* to reflect the pervasive use of electronic display panels (McClumpha, James, Green, & Belyavin, 1991). All new commercial airliners are dominated by electronic displays that are indications of the extensive computer processes behind the scenes that intervene between the raw data received from sensors and the presentation made to pilots. This has the side effect of divorcing the crew from the raw data unless that data happens to coincide with a specific need of the pilot for the information. More often, the result is processed data that is intended to make the pilots interpretation of the data quicker and easier (McClumpha et al., 1991). The net effect of the increase in processing power has been to simultaneously increase the level of system autonomy, authority, complexity, and coupling while changing the role of the human crewmember to more of a system monitor, exception handler, and manager of automated resources (Sarter, Woods, & Billings, 1997; Spirkovska, 2004). Policy on the use of cockpit automation has not always been standard. Although the FAA dictates some minimum standards in their regulatory guidance, aircraft operators have a great deal of flexibility on the policies that their crews follow. These policies run across the spectrum from requiring use of all automation available, to minimal use necessary with the individual flight crew making the determination (Young, Fanjoy, & Suckow, 2006).

The evolution to the current level of automated cockpit systems has provided benefits in terms of flying safety, efficiency in the use of limited airspace traffic flow, and economy of flight to both airlines and consumers. Aviation is considered one of the safest methods of transportation with an accident rate of less than two per million departures, automated flight computers have allowed more aircraft to use the same airspace in a given period of time, and computerization has allowed the elimination of navigators and flight engineers, reducing training and personnel costs (Olson, 2000). However, automation has not been without negative effects. The changing role of the pilot has introduced new modes of failure, redistributed workloads such that it is—at times—higher than before at critical phases of flight, and increased training requirements (Ancel & Shih, 2014). Contrary to the implication of the term automation, humans are still critical to the system. The human role has now changed to a manager and a monitor who looks for failures and for conditions that the automation has not been designed to handle. This is not a role people are ideally suited to accomplish (Endsley, 1996).

Not surprisingly, the new demands on the pilot have resulted in calls from many quarters for more guidance on both the design of the systems, and the requirements and limitations on aircrew members for controlling and monitoring the automation. Historically, the FAA has viewed their regulatory role in this ecosystem as setting minimum standards for pilot qualification and for equipment certification, preferring to leave the detailed guidance to the operator. U.S. airlines had looked at the result of accidents such as the Cali crash and came to the conclusion that the training they have provided to crews at the time had not been up to the task (Amalberti, 1999). However, the increasing number of aircraft mishaps being attributed to automation and automation-related human confusion has not led to any universal guidance on what is appropriate or not for how automation is applied and used (Spencer Jr., 2000).

A considerable number of studies have already been carried out which look at specific effects of human factor issues in aviation, such as workload distribution or conflicts (Hilburn, Bakker, Pekela, & Parasuraman, 1997; Hollnagel, 2007), the increased risk of collisions (Hoekstra, Ruigrok, & Van Gent, 2001; Hollnagel, 2007), or contribution to the loss of control (Geiselman, Johnson, & Buck, 2013; Lowy, 2011). These have all looked at the role of the individual in the chain of events leading to aircraft mishaps. Numerous researchers have identified individual issues of mode confusion (Faulkner, 2003; Rushby, 2001, 2002; Silva & Hansman, 2015; Young et al., 2006), loss of situational awareness (Parasuraman & Riley, 1997; Spirkovska, 2004; Young et al., 2006), poor training (Joslin, 2013; Sarter et al., 1997), and inadequate human-machine interfaces (Geiselman, Johnson, & Buck, 2013; Sarter & Woods, 1994) that could address the last point of failure before some tragic event occurs: the pilot. What is missing and necessary for continuous improvement in the long-term safety of air travel, is a method to look beyond the physical act (or lack of action) that caused the mishap to occur, to the underlying factors that led to that act, including those attributable to the government, the airline, the supervisor, and the crew composition. In the flying world, mishaps may metaphorically be thought of in terms of a chain of events in which any action to break that chain would have prevented the final outcome.

Purpose

This study examines both quantitative and qualitative data from the United States Government to determine the impact of policy on aircraft mishap rates and pilot behavior. Data from the National Transportation Safety Board (NTSB) and the Aviation Safety Reporting System (ASRS) provide a detailed insight into both expert-evaluated data from formal investigations and first-hand information from pilots involved in mishaps. NTSB investigations systematically document all actions taken and contextually relevant facts (e.g. weather, other traffic, etc.), and then uses acknowledged experts to make informed decisions on the likely causes as well as recommended courses of action to prevent recurrence of the identified problems (National Transportation Safety Board, n.d.-c). The ASRS is an incentivized program for pilots to report mishaps and near-mishaps in a protected environment to advance the safety of flight (National Aeronautics and Space Administration, n.d.). The unique combination of first-hand source material from pilots and the filtered material of expert analysts provides a clear insight into the relationship of regulatory guidance, cockpit automation, and safety.

Overall, this study will examine the data from NTSB and ASRS reports and investigations with specific regard to the effect of automation and of policy guidance in place at the time of the mishap. Specifically, the research questions guiding this study include:

1) Does the implementation of automation technology have a significant impact on the incidence of aircraft mishaps in United States Aircraft?

   a. If there is an impact, what is the impact of human factor issues?

   b. If there is an impact, what is the impact of technological deficiencies?

   c. If there is an impact, what is the impact of human-machine interfaces?

2) Is the current policy guidance on cockpit automation sufficient with respect to:

   a. The scope of guidance provided to passenger carrying aircraft?

   b. The relevance of guidance to the current state of technology?

   c. The compliance and enforcement of such guidance on airlines?

3) Does the type of training received by cockpit crew and the level of experience with cockpit automation change the relationship of flight deck automation and aircraft mishaps?

4) What are the policy actions that have shown an ability to mitigate negative effects or enhance positive effects of flight deck automation on aviation safety?

The sample used to determine the impact of automation will include all reports from 1996 to the present because the Cali, Colombia accident in December 1995 represents a clearly established initiation point for identifying automation as a primary cause or significant contributor to the accident (Lintern, 2000).

The analysis of these questions will be performed through an explanatory sequential mixed methods approach. The sufficiency of existing guidance was assessed quantitatively via an ordinal regression analysis that examined the dependent measure of mishap severity—classified as minor, serious, and fatal—against the independent variables of whether various types of automation were found in the causal analysis determine the significant factors that can be addressed by policy. To analyze the effect of technology, human interaction with technology, and training on mishaps, as well as to evaluate the efficacy of various policies, a qualitative approach explored the causal analysis reports from the NTSB for various mishaps during periods of variation in those factors.

Theoretical Framework

One of the more recent mishaps in which inappropriate use of automation was determined to be a factor occurred on July 6, 2013 at San Francisco International Airport. Asiana Airlines flight 214 was on final approach to land after departing from Incheon, Korea and hit the seawall at the end of the runway, breaking the airplane into pieces, leaking oil onto the engine and creating a large fire. There were 307 passengers on board and three people suffered fatal injuries. The flight before the landing phase was uneventful. On approach, the cockpit clearly displayed an indicated airspeed of 103 knots (118 mph) when the appropriate airspeed should have been 137 knots (160 mph). According to experts, this should have been a clear indication to break off the approach and reattempt (Vartabedian, Weikel, & Nelson, 2013). Interviews with the pilot make clear that he believed the auto throttles were engaged and the airplane would maintain the 137 knots required. There was no indication of any mechanical failure (Chow, Yortsos, & Meshkati, 2014).

Following this tragic event, the NTSB performed an investigation as their charter dictates. At the opening of the NTSB hearing, the acting chairman, Christopher Hart quoted the noted psychologist James Reason when he said "in their efforts to compensate for the unreliability of human performance, the designers of control systems have unwittingly created opportunities for new error types that can be even more serious than those they were seeking to avoid" (Hart, 2014, p. 20). It was fitting that the chairman would choose to refer to Reason as his 1990 work, *Human Error,* is recognized as a seminal work on the theory behind how human errors promulgate in complex systems to result in significant or catastrophic failure (Reason, 1990). Prior to Reason's work, the construction of a proper systemic model of mishaps across time and geography had proved difficult to develop (Adachi, Ushio, & Ukawa, 2006; Norman, 1988).

The model he developed is called the Cumulative Effect Theory, but is best known by practitioners as the Swiss Cheese model for accident causation (Koeppen, 2012). Reason (1990) was the first to move past the precedent of focusing on active human errors and analyze latent errors which may lie dormant in the system—and outside the individual's control—to break through all the defenses the system has attempted to erect. Moving beyond the most immediate cause, which is usually attributed to some kind of pilot error, and finding latent weaknesses that can be addressed to prevent future accidents is critical in the current environment of rapid technological development and change. As Reason (1990) notes, errors in these latent layers "pose the greatest threat to the safety of a complex system" (p. 173).

In this model, shown in Figure 1, failures can be influenced at multiple levels of the process. At each level of possible intervention, there is a potential for process failure which is analogous to slices of Swiss cheese stacked together. At each level, there are holes—or potentials for allowing errors—which may allow a problem to pass through a

hole in that layer. However, the next layer has holes in different places and the problem

should be caught. Each layer represents a defense against any potential error impacting

the final outcome. For a catastrophic error to occur, all the holes in each defensive layer

need to align in order to overcome the barriers as shown in Figure 2. If the holes in each

layer are set up to align, then the system is inherently flawed and will allow negative

outcomes (Reason, 1990).

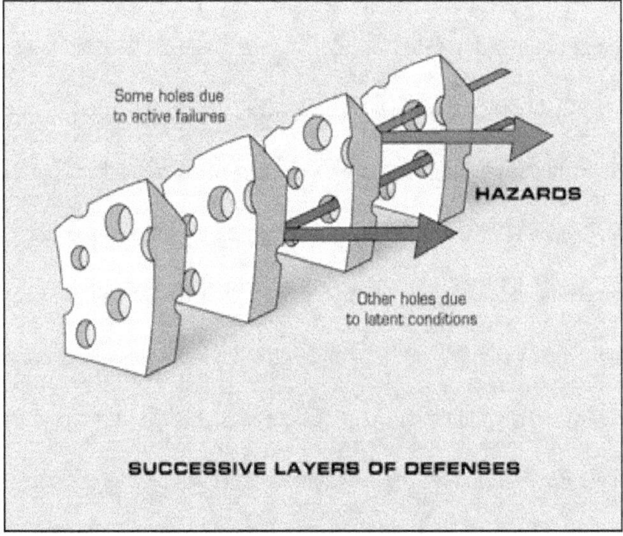

Figure 1. Cumulative Effect Theory or the Swiss Cheese model (Reason, 1990), graphic modified from *Anatomy of an Error* (Duke University School of Medicine, 2016).

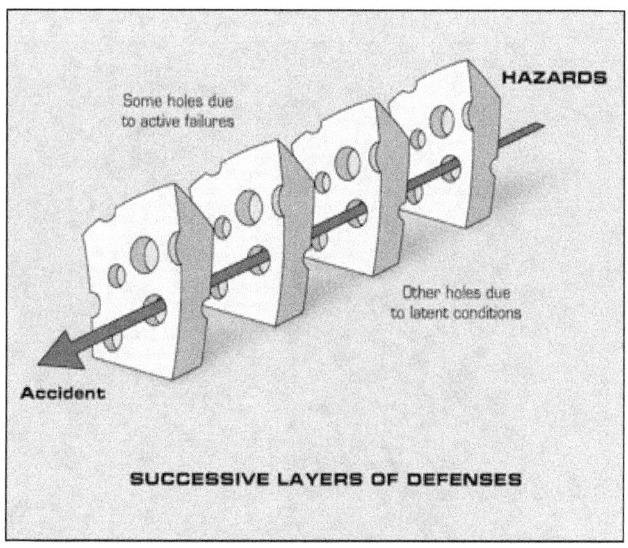

Figure 2. Failure mode in the Swiss Cheese Model (Reason, 1990), graphic modified from Anatomy of an Error (Duke University School of Medicine, 2016).

Shappell and Wiegmann (2000) adapted Reason's original model with a view toward aircraft accident causation as shown in Figure 3. "The Human Factors Analysis Classification System (HFACS) is a general human error framework originally developed and tested by the U.S. military as a tool for investigating and analyzing the human causes of aviation accidents" (Wiegmann & Shappell, 2001, p. 1). The HFACS describes the holes in each layer of the model that allow aircraft accidents to occur and defines each of the four layers as tiered systems that affect the layer below: organizational influences, unsafe supervision, preconditions for unsafe acts, and the unsafe acts. Analyzing an accident typically starts at the most obvious layer—the act—and works backwards to identify latent causes (Shappell & Wiegmann, 2000).

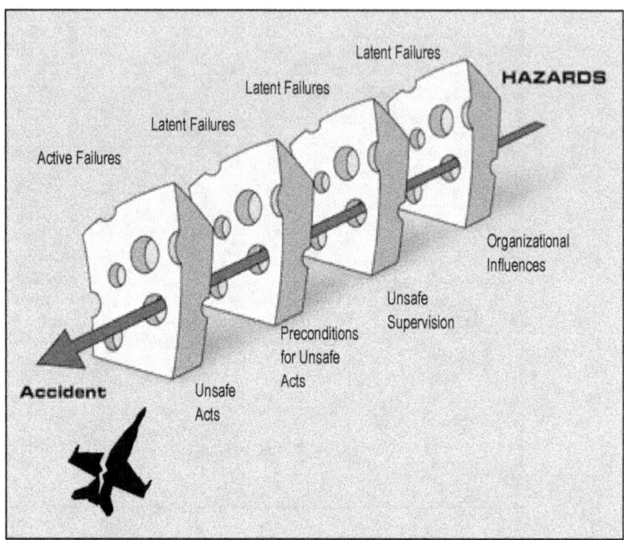

Figure 3. Shappell and Wiegmann's adaptation of Reason's model for aircraft accident causation (Shappell & Wiegmann, 2000) graphic modified from Anatomy of an Error (Duke University School of Medicine, 2016).

The first layer refers to the actual unsafe acts, often called pilot error and shown in Figure 4. This level encompasses both human errors and intentional violations. Errors are unintentional and can be due to faulty perception, inadequate skill or training, or mental mistakes. Violations are intentional and can be routine in that the pilot perceives the benefit of the violation to outweigh the risk, or exceptional such as in the case of an emergency procedure requiring a deviation from approved procedures (Koeppen, 2012).

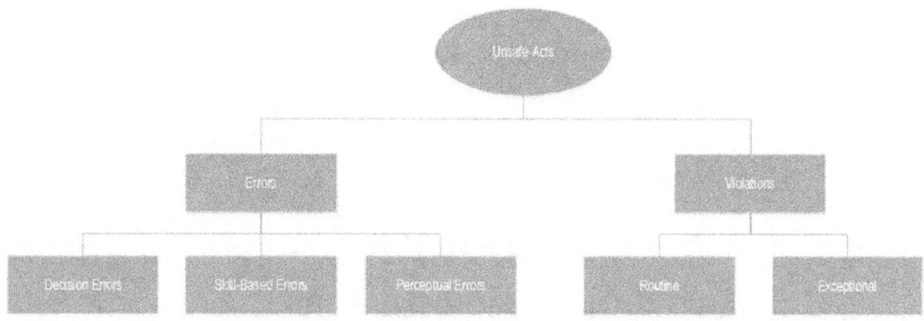

Figure 4. Shappell and Wiegmann's analysis of unsafe acts (Shappell & Wiegmann, 2000).

"Arguably, the unsafe acts of pilots can be directly linked to nearly 80% of all aviation accidents" (Shappell & Wiegmann, 2000, p. 6). However, the benefit of the Swiss Cheese model lies in the ability to go beyond the act and identify latent failures present in the system that allow the unsafe acts to take place. Therefore, the preconditions that allow for the unsafe acts are analyzed. This includes factors such as fatigue, poor communication, failures in coordinating procedures, or failures in crew resource management (CRM). These are organized into the broad categories of substandard condition of the operators or substandard practice of the operators as shown in Figure 5.

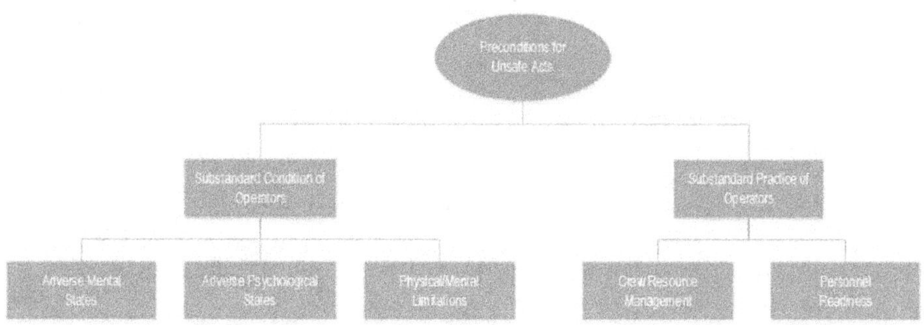

Figure 5. Shappell and Wiegmann's analysis of preconditions for unsafe acts (Shappell & Wiegmann, 2000).

Reason (1990) linked causal factors associated with pilot error back to errors made in the supervisory chain. Hence, Shappell and Wiegmann (2000) identified four categories of unsafe supervision. The categories include violations by the supervisor, failure of the supervisor to correct problems, operations that were inappropriate from the original plan development, and inadequate supervision as shown in Figure 6. These categories cover a range of failures on the part of supervisors that range from failure to provide policy or doctrine, poor training, improper manning or crew rest control, allowing documentation errors to go uncorrected, failure to initiate corrective actions against substandard crewmembers, or failure to enforce existing rules and regulations (Shappell & Wiegmann, 2000).

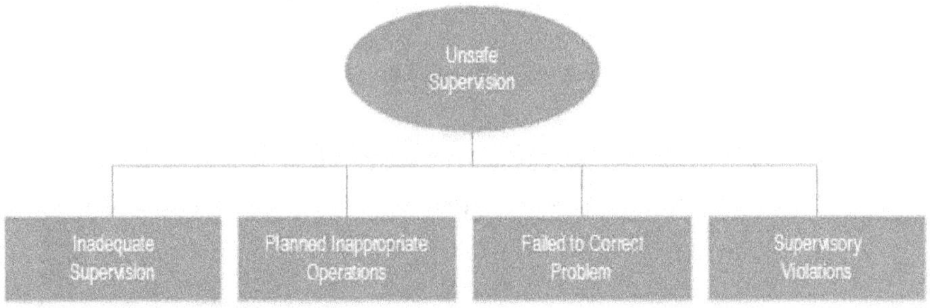

Figure 6. Shappell and Wiegmann's analysis of unsafe supervision (Shappell & Wiegmann, 2000).

Organization influences covers the failures within the organization that may include items such as resource management, climate, training, experience, and process control and are shown in Figure 7. "Unfortunately, these organizational errors often go unnoticed by safety professionals, due in large part to the lack of a clear framework for which to investigate them" (Shappell & Wiegmann, 2000, p. 11). Wiegmann & Shappell (2001) described these as being the most elusive of latent failures. Examples of these failures can include human resource or budget shortfalls, organizational structure and oversight, and published operating procedures (Koeppen, 2012; Olson, 2000).

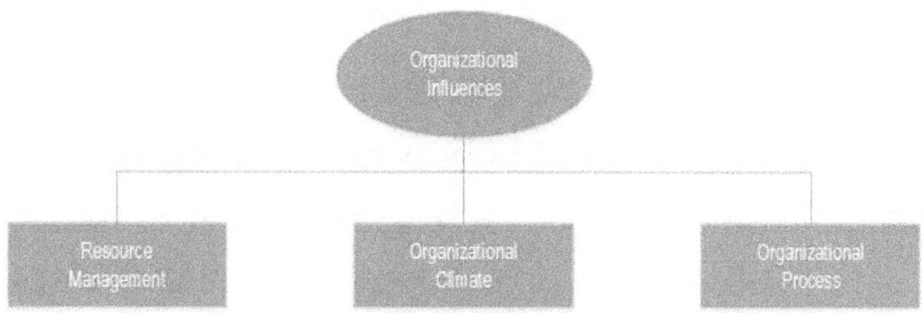

Figure 7. Shappell and Wiegmann's analysis of organizational influences (Shappell & Wiegmann, 2000).

Influence of Policy on Aircraft Automation

Since the mid-1990s, commercial aircraft fatal accidents have decreased by nearly 80% while simultaneously since 2000 more than 16 large and medium airports have opened with the capacity to accommodate more than 2 million additional operations (Federal Aviation Agency, 2011). This incredible rate of growth coupled with achieved safety levels is unprecedented and exceeds the goal published by ICAO (Sherry & Mauro, 2014b). ICAO recognized automation as a significant contributing factor to this combination of safety and efficiency (Federal Aviation Agency, 2011). However the safety rate has recently leveled off, and to continue improvement in safety as air travel grows requires new methods of analyzing the system. The human versus automation dynamic has changed. The human pilot is much more a supervisor than an executor, while the automation has taken previous roles from the pilot without taking associated responsibility (Geiselman, Johnson, & Buck, 2013). Automation has design limits and it requires the human operator role to shift toward more of the cognitive processes of judgment and decision making in order to ensure proper operation at the most critical times.

The FAA is currently in the midst of a massive overhaul of the way aviation operations are conducted in the United States. Their NextGen program aims to completely transform the National Airspace System (NAS) in terms of structure, operating procedures, and assignment of responsibilities. This modernization program will decentralize decision making from air traffic controllers to individual flight crews, increase the dependence on technological advances to reduce aircraft separation criteria, and increase volume of air traffic in already congested airspace (Federal Aviation Agency, 2009). The introduction of this new slate of capabilities and responsibilities to the existing system offers the possibility of increased risk to public safety. Historically, the imposition of new policies and procedures in aviation has taken a retroactive stance, in which policy is written in response to some negative action that has garnered attention (Sawyer, Berry, & Blanding, 2011).

CHAPTER II

LITERATURE REVIEW

In the analysis of modern aircraft flight decks, the term automation could be ambiguous. In the context of aviation policy, the term is understood to mean the various forms of technology that interface with the flight control and navigation systems that automate execution of flight tasks by the use of self-operating machines or electronic devices (Koeppen, 2012). Although processes may be automated by the systems installed on the airplane, they do not operate completely autonomously and the pilots are still central to all operations and the safety of the flight. Billings (1991) expanded on this to consider the unique nature of the human-machine interaction by highlighting the role of the automation as a tool or resource available to the human designed to facilitate the accomplishment of required flight tasks with greater efficiency.

The transition from pilot seat-of-the-pants flying with limited outside help to the technological and operational implementation of automated systems started taking shape in the 1980s (Salas, Jentsch, & Maurino, 2010). As the efficiency value of automation was recognized, little attention was paid to the impact of automation on crew performance. However, since the advent of technology based automation, these systems have altered the role of the pilot and become more complex. The need for pilot assistance has been recognized since the beginning of aviation, even prior to the technological capability to provide such aid, because "not all of the functions required for mission accomplishment in today's complex aircraft are within the capabilities of the unaided human operator" (Billings, 1991, p. 8). A parallel desire for accomplishing more complex missions was for improving safety and increasing the efficiency and capacity of an increasingly crowded United States NAS (Thomas & Rantanen, 2006).

The first technological automation for cockpits was originally developed with the hope of meeting those missions by increasing precision, reducing workload (Sarter et al., 1997), improving economy and efficiency, and reducing manning requirements while simultaneously improving safety (Young et al., 2006). Developers originally considered it possible to create autonomous systems that required very little or even zero human involvement in order to eliminate the impact of human error. The engineering view was that automation could be substituted for human action without any larger impact on the system other than the output being more precise. This view was predicated on the assumption that the complex tasks associated with flying and navigating a commercial airplane could be decomposed into essentially independent tasks without consideration of the human element in the overall system (Sarter et al., 1997).

The current state of the art among airlines includes numerous examples of automation technology that are considered indispensable for modern flight operations and are shown in Appendix B. The FMS supports flight planning, navigation guidance, performance management, automated flight-path control, and monitoring of flight progress (Sarter & Woods, 1994). Commercial airlines—which operate under the auspices of the Part 121 of Section 14 of the CFR—for many years have been using Electronic Flight Bags (EFBs) to compute flight performance, present navigational charts, and display instrument approach charts in order to direct aircraft to a safe landing under adverse weather conditions (Chandra & Kendra, 2010). Cutting edge developments in technology often manifest themselves as new features available in the EFB, including weather forecasts and radar displays, maintenance documentation, engine and airframe health status, and voice/data communication (Chow et al., 2014). Airline confidence in the efficiency and safety of modern systems, coupled with the cost savings associated with their use—efficiency results in reduced use of fuel, crew training, and pilot manning requirements—have led many companies to mandate the fullest use of automated systems (Young et al., 2006).

Automation Functions and Human Effects

In practice, the effort to replace human intervention has proved problematic and controversial. Although the superior computational capacity and reduced reaction time of technology has been beneficial, replacing humans can increase system vulnerability with regard to unanticipated perturbations that engineers have not designed into the software (Dalcher, 2007). In practice, the effort invested by air carriers in training, coupled with the experience and expertise of individual pilots usually compensates for the features of automation that have proved detrimental (Woods & Sarter, 2000). Systems on commercial airliners that are designated as critical to safety of flight are certified to have a failure probability of 1×10^{-9} (Sherry & Mauro, 2014a). Operator training and expertise has resulted in the outstanding safety record demonstrated in the NAS despite any clumsiness or limitations in system designs. But as technology gets more complex and automation takes a larger role, human factor concerns have been raised with regard to loss of manual flying skills, reduced situational awareness, mode confusion, and inadequate feedback (Young et al., 2006).

For the most part, it is fairly clear that automation has worked quite well and resulted in many of the desired goals when first introduced (Wiener, 1985), however there has been a great deal of empirical evidence of negative effects (Martinez, 2015). Negative effects can reasonably be predicted because whenever a new technological system is introduced, something will always operate differently than was planned or expected by the designers, and the operators need time to adapt the system to the way they actually use it and to adapt their existing processes to overcome quirks in the design (Hollnagel, 2007).

The FAA, NTSB, aircraft manufacturers, and academic researchers have made an effort to examine the safety impact of automation. In general, the results of this have suggested that offsetting the potential benefits to pilots and the airspace system in general, there is an increase in the chance of accidents directly related to the systems themselves or the pilots training in their use (Franza & Fanjoy, 2012). This finding is in accordance with the existing concept of operations for flight deck automation in which the crew delegates routine tasks to automation components and then supervises performance of those components. The crew is only expected to intervene when an inappropriate command or output is observed. Because such events are rare in practice, pilot ability to counteract them is often compromised by two different factors: (a) lack of knowledge of the system's inputs, logic, and prioritization; and (b) poor communication of the problem by the system and the corrective course of action to the crew (Sherry & Mauro, 2014b).

Modern automation technology is inherently complex in order to deal with the complex operating environment in which it is deployed. The system is constructed by teams of engineers—often not pilots themselves—distributed in groups with great diversity of geographic location, cultural experiences, technological skills sets, and understanding of the larger strategic system. It is not feasible to expect pilots to understand the complete behavior of such systems, especially for aspects of the technology that occur in the programming that is not visible to the user (Sherry & Mauro, 2014b). Addressing such human-machine interface issues must be the responsibility of the designers as part of their development process. However, accident investigations (see Appendix A) regularly show that these concerns have lagged the development of the technology itself. Inability of a system to communicate its state to the crew, failure to provide the pilots with the context necessary for them to maintain a state of situational awareness, display modes that are easily overlooked during times of heavy workload, and instigation of unexpected flight inputs that the crew are not prepared for are contributing factors to aircraft mishaps often cited in final reports by investigators (Parasuraman & Riley, 1997).

As humans living in the modern world, we have first-hand experiences daily of technology not acting in the way we would expect everywhere from home to work to our vehicles. It is therefore not unexpected that aviation technology would have similar imperfections (Thomas & Rantanen, 2006). Accepting this premise dictates the conclusion that human intervention will be needed at some point to correct for a situation not adequately planned for in design. Making the human flight crew responsible for this new supervisory role requires a social change from the pilot being the sole or final arbiter of flight control inputs to a new paradigm where his or her input is normally either not needed at all or only entered as a result of instructions received from an on-board computer. Pilots—through their unions—initially resisted this change. Even though the current generation of airline pilots has come to accept their increased role as computer programmer, automation monitor, and emergency override, that role is still at odds with the training that prepared them to be pilots in the first place. As with previous implementations of new aircraft technology, a period of adaptation is needed. It was the case when autopilots were introduced, when jets replaced propellers, when hydraulic controls replaced manual cable control, when computers were interjected into black boxes controlling instrumentation, and is now the case as automation takes a dominant role (Amalberti, 1999).

Human-Automation Interaction

As Jordan (1963) noted over 50 years ago, the proper role for humans and machines interacting requires viewing them as complementary rather than independent. They must work together to achieve the design level of performance. Regardless of the level of automation, there will always be a need for the presence of a human operator, even if the need is only to intervene when there are system abnormalities or emergencies. This ongoing requirement for human-automation interaction has changed the role of the modern pilot significantly.

Sarter (1997) highlighted the role of the pilot as both a translator of air traffic control clearances into commands the automation can understand, and mediator for reasonableness of the clearance to the system. The role also includes a responsibility to determine when the automation needs to be directed to invoke a new mode or course of action. These pilot actions typically occur simultaneously with the increase in cockpit workload such as changing flight levels, preparing to begin an instrument approach, changing radio frequencies, or updating navigation and weather information. Because of the inappropriate timing imposed by the automation requirement, Wiener (1985) coined the phrase *clumsy automation* to capture the concept of automation causing a redistribution of workload over time rather than changing the amount of work—up or down—with the new added roles of the pilot.

A widely accepted taxonomy for the new role of aircrew members has been to break down the actions of the pilots into those of planning, teaching, monitoring, intervening, and learning as shown in Table 1 (Olson, 2000). In the role of planner, the pilot decides what inputs need to be made, which variables to manipulate, and the criteria for which actions the automation is to take control over. As a teacher, the pilot instructs the automated system as to the appropriate targets and algorithmic instructions. The pilot must then monitor the system to ensure it behaves as expected and is free from malfunctions. When there are deviations or fault detections, the pilot must then not only intervene, but intervene within an appropriate time frame and with the appropriate corrective action. Finally, the crew must synthesize all that has happened on a particular flight to learn lessons that may be applied to future system control situations. Each of these steps of human interaction is susceptible to errors.

Table 1

Human Roles and Opportunities for Error in Supervisory Control

Human Role	General Difficulty	Caused by	Contributing
Planning	Inappropriate plan developed	Failure to consider relevant information	Inadequate mental model
			Inert knowledge
		Failure to understand automated system	Mode errors
Teaching	Improper data entry	Wrong data/incorrect Location	Time delays
Monitoring	Failure to detect the need to intervene	Human monitoring limits	Inadequate mental models
		Expectation based monitoring	Information overload
		Inadequate feedback	Lack of salient indications
			Keyhole property of FMC
Intervening	Missed/incorrect intervention in undesired system behavior	Inability to understand: Why the problem occurred? What to do to correct it?	Inadequate mental models
			Complex systems
Learning	Failure to learn from experiences		Inadequate mental model

Note: Taken from Olson, W. A., 2000, *Identifying and mitigating the risks of cockpit automation*, p. 12.

Aircraft mishaps are often attributed to pilot error, and the evaluation of accident investigation experts should properly be given appropriate weight. However, to examine only the role and mistakes of the pilot when looking at failures of automation to perform properly is to miss the strategic opportunity to reduce the prevalence of human error by addressing the human-machine interface (Geiselman, Johnson, & Buck, 2013). Many modern systems have modes that are only used by some, but not all, operators. Pilots are not trained for these modes if their carrier does not approve usage, but their existence provides additional complexity when seeking information or programming actions (Woods & Sarter, 2000), leaving many commercial pilots baffled in some situations while trying to understand what the automation is doing, why it is taking a certain action, and what action it will direct next (Klien, Woods, Bradshaw, Hoffman, & Feltovich, 2004).

This result points to issues of crew training as well as cultural issues. As aviation consultant Jay Joseph was quoted in the *San Jose Mercury News*, "some of the problems we have now are with a new generation of pilots who are accustomed to everything being provided to them electronically, even to flying with the autopilot. It's a sad commentary about where aviation has gotten itself" (Nakaso & Carey, 2013, p. 8). The consequences of flight deck automation have generated at least two unanticipated side effects for the pilots: the first is that the consequence of an error is often shifted to some point in the future when the automation acts on the incorrect input, and the second is that feedback designed to aid the pilot can often turn into a trap as the pilot puts too much reliance on that feedback alone at the expense of other flight instruments (Amalberti, 1999). As long as humans are involved in the process—whether in the design, manufacture, or execution—there will always be mistakes. It is simply impossible to anticipate and deal with every combination of unexpected or random issues that can be encountered or the human limitations of forgetfulness, tiredness, inattention, or other factors (Maurino, Reasonson, Johnstonton, & Lee, 1995).

One area of improvement critical to minimize the effect of human error with regards to cockpit automation is to improve the human-machine interface in the areas of feedback and communication. The modernized system needs:

- To indicate when the automation is having trouble handling the situation

- To indicate when the automation is taking extreme action or is moving toward the limits of its authority

- To consider the nature of training required

- To have human factors considered as part of the design

- To have the authority of the automation appropriately limited. (Sarter et al., 1997, p. 8)

The scope of authority of the system to act independently is of significant concern and is a source of disagreement among researchers and practitioners. Jordan (1963) says "we can never assign [the machines] any responsibility for getting the task done; responsibility can be assigned to man only" (p. 164). The point being that no matter how advanced our technology may have become, there is a key difference in human-human interaction compared to human-machine interaction. Specifically, we may treat an intelligent machine as a rational being, but not as a responsible being (Sarter et al., 1997).

This differentiation is built into the modern flight deck system, in which flight crews delegate tasks to the automation and monitor its performance. In the event of inappropriate automation commands, it is the responsibility of the crew to intervene and take corrective action. However, the ability of the crew to do so is limited by several inherent design features of the automation systems:

- The hidden (from pilot view) nature of the fail-safe sensor logic

- Silent or masked automation responses

- The absence of cues to anticipate performance envelope violations

- Difficulty in recognizing performance envelope violations due to extraneous notifications

- Non-linearity and latency of aircraft performance parameters that make it difficult to recognize violations without assistance. (Sherry & Mauro, 2014a)

When automation is granted a high level of authority over the functions and control inputs of an aircraft, the pilot will require a proportionally higher level of feedback than if he or she were making the inputs themselves so that they can monitor whether the intended actions match the actual ones (Parasuraman & Riley, 1997). Unfortunately, multiple studies have shown that current automation does not provide such feedback. Visual or audible alerting features were often missing or confusing, leading to misdiagnosis of a problem or complacency (Spencer Jr., 2000). These limitations imply that—no matter the sophistication of the computer technology involved—automated systems cannot be trusted to share responsibility for safety of the aircraft with the human. Pilots must retain primary responsibility for detecting and resolving conflicts between human goals (flight path adherence, mid-air avoidance, etc.) and machine actions (climbs, descents, turns, etc.) (Olson, 2000).

Automation can be a powerful tool in avoiding and compensating for trouble, but the problems that have been identified highlight the need to ensure that humans and automation interact in an efficient manner. This requires that designers spend a great deal of effort in planning machine actions and feedback in a way that highlights the appropriate areas of interest to the pilot at the appropriate time and in an easily digested format. The area of work that concentrates on these considerations is human factors. Human factors examines the concepts, principles, and methods necessary to meet objective criteria for the scientifically based rules of human sensory, perceptual, motor performance, and cognitive performance to maximize efficiency of the pilot interface environment (McClumpha & Rudisill, 2000).

Human Factors Considerations

When evaluating the human factors in a complex system, the number of variables and variations can become overwhelming. Compounding that complexity is the fact that often the variables in question have interactions on multiple levels and react differently when in the presence or absence of yet another variable. The cost and time required for empirical research on all possible implications is prohibitive and would require very large experimental designs that would still suffer from lack of a control group to ensure scientific levels of validity and statistical power (Thomas & Rantanen, 2006). Because of these limitations, aircraft research and design in the area of human factors has traditionally focused on those areas identified by aviation experts that have been exposed by real-life experience or accident investigation findings, such as situation awareness.

The aircraft cockpit is a socio-technical system and as such, the human's place and role in that system requires specific attention. According to the ICAO one of the main human factor considerations is situation (also called situational) awareness

(International Civil Aviation Organization, 2013). This can be defined as perceiving, comprehending, and forecasting the state and position of (a) the aircraft and its systems; (b) location of the aircraft in four dimensions; (c) time and fuel state; (d) potential threats or dangers to safety; (e) contingencies and what-if scenarios for potential actions; and (f) awareness of the people involved in the systems including other crewmembers, outside agencies, and passengers.

Designing automation systems that account for human factors such as situation awareness is a difficult challenge. For the designer, predicting human intentions by only monitoring their inputs to an automated system is problematic, knowing what feedback the pilot requires at what time is ambiguous in the sanitary environment of the design lab, and liability issues impose even more restrictions on the ability of technology to infer human planning (Inagaki, 2003). There are many points of data that a pilot needs to know or needs ready access to, and they change over time with the phase of flight being encountered. Pilots must understand the flight path, the equipment requirements for any individual maneuver being performed, the terminology of various types of maneuvers identified by ICAO which change over time, air traffic control intentions and terminology, engine and flight control status, fuel state, weather situation, and anticipated changes in all of these items over time (Chandra et al., 2012). Poor situation awareness can also lead to other problems in using or controlling automated systems. Such problems may include mode error (i.e. pilot actions inappropriate for the given aircraft mode due to the pilot believing they are in a different mode) or out of the loop error (i.e. the pilot not understanding what actions the automation has directed and therefore not being ready to take over control from the automation) (Olson, 2000).

The types of controls and warnings needed to improve pilot situation awareness must indicate when the automation is having trouble handling a situation, when the automation is taking an extreme action or is operating at the margins of its authority, and when the agents (pilot and machine) are in competition for control (Woods & Sarter, 2000). Most modern accidents are viewed as a result of a system failure rather than a failure of one single component of that system, so preventive actions need to focus on the interactions between system components such as between human and machine (Vuorio et al., 2014). The complexity of those interactions can be simplified via better feedback, more practice or training, teaming up the machine with the human for actions, and creating intuitive designs that are learned quickly and can be operated simply during times of stress. The final outcome of system interactions is a much more important result of human factor considerations than the behind-the-scenes programming of any automation (Thomas & Rantanen, 2006).

Pilots have been shown to utilize a risk-time model to categorize information they receive during flight in which they only consider additional information and options if time allows. Under this model—trained from the earliest days of pilot training—safety is the first consideration. The universal flight training axiom is to aviate, then navigate, then communicate, meaning fly the plane safely first, then determine where to go and how to get there, and finally tell outside agencies of your plan (Spencer Jr., 2000). In order to meet the demand of this internal mental model, pilots are often inclined to use the most salient source of information available at any given time. This often turns out to be the automated indication, even at the expense of more accurate, but difficult to access information located elsewhere. Additionally, company policies may promote using a specific data source at the expense of other feedback systems (Mosier, Sethi, McCauley, Khoo, & Orasanu, 2007). These pilot models need to be considered in designing automated feedback systems.

Implementing a proper human factor consideration requires looking beyond the current functions being modeled or replaced by automation. Automated devices also often create new demands on the individuals using them. There can be new tasks, changed tasks, new cognitive demands, new knowledge requirements, new forms and requirements for communication, new types of data to be managed, additional requirements for a pilot's attention, changes in the timing and time required for previously existing actions, and new forms of error for which a pilot needs to account (Sarter & Woods, 1994). Humans are inherently somewhat unpredictable in how they respond to experiences for which they have not had prior exposure. Technology that replaces human actions will inevitably present a pilot with a unique situation and hence, the predicted action of the pilot is difficult to foresee (Dalcher, 2007).

The most common way to address human factor considerations to date has been through increased training in one of two areas. The first area is in training on the automation itself. Authorities and airlines alike have asked for a more procedure-driven approach for which standards and training can be developed to minimize pilot errors (Amalberti, 1999). This kind of training focuses on what is controlled in each mode of automation, where each mode gets the data it needs to make decisions and take action, where each mode obtains its target for results, and what action will each mode take once the target result is achieved. Furthermore, this type of training emphasizes the role of the pilot in finding the relevant information, attending to the sources of information, interpreting the information correctly so as to monitor the systems, and integrating this new information into their existing knowledge base on the aircraft and its intended flight path (Sherry & Mauro, 2014b). However the results of this kind of training are questioned when taken in the context of unpredicted outputs (Koeppen, 2012) or when economic pressures limit the amount and duration of training outside the profit making activities of pilots (Amalberti, 1999). Studies have shown that airline-developed training and published standard operating procedures that focus on the "why" of actions rather than the technical explanation of "how" are somewhat more successful in achieving pilot compliance with their training in actual flight conditions (Giles, 2013).

Yet, the perceived or actual shortcomings of technical training led to a second, more comprehensive training approach, originally called cockpit resource management, and now called crew resource management, but known by the acronym CRM in 1979. CRM was first defined as the set of teamwork competencies that allow a crew to handle situational demands that might normally overwhelm an individual crewmember (Koeppen, 2012). This definition was expanded over the years and refined in FAA guidance that requires CRM training to focus on standard operating procedures (SOPs), the functioning of crewmembers as a team rather than a collection of technically competing individuals, methods of behavior that foster crew effectiveness, providing opportunities for crewmembers to practice the skills necessary to be effective team members and team leaders, and appropriate behaviors for crewmembers in both normal and contingency operating situations (Martinez, 2015). During the career of a pilot, that individual will experience multiple opportunities for CRM when they are first hired, when they transition to a new model of aircraft, when they upgrade to a new crew position, and during annual proficiency training. The intent of the repetition is to create habitual behavior that reinforces efficient crew and automation interactions (Giles, 2013). While training in CRM has shown success in reducing human errors, it is limited by lack of standardization across airlines and with the difficulties in reacting to some of the problems created by lack of proper human factor analysis (Hendrickson, 2009).

Automation-Induced Problems

One of the most significant problems to which excessive reliance on automation may contribute is the loss of situational awareness, because it may lead to several other follow-on problems. As Endsley (1996) says, "situational awareness (SA) can be conceived of as the pilot's internal model of the world around him at any point in time" (p. 97). A lack of situational awareness by pilots can lead to catastrophic or fatal system failures (Koeppen, 2012). Automation can contribute to loss of situational awareness in several ways: (a) being overly attentive to automated flight modes can cause loss of awareness of basic flight parameters, (b) being uncomfortable with computers may cause pilots to defer their use to other crewmembers and damage crew coordination, (c) overconfidence in automation can lead to a passive role by the pilot in which they lose the mental schematic of what is going on around them, and (d) alerts—whether normal or abnormal—can distract a crew from other priorities (Endsley, 1996; Young et al., 2006). Many of the factors that cause detrimental impacts on situational awareness can be traced back to the system design. These factors can include poor feedback mechanisms, limited training resources, overly autonomous systems in which the technology directs action without knowledge of the pilot, and highly interdependent systems that cause unexpected flight control inputs in systems other than the one making a decision (Sarter, Mumaw, & Wickens, 2007). Therefore, it is essential to minimize potential problems in the design phase before the automated system goes into operational use to realize the intended benefit without depriving the crewmember of her or his required situational awareness (Endsley, 1996).

One of the second-order effects of losing situational awareness is mode confusion. Cognitive scientists believe that humans construct mental models of the world and then use that model to guide their interactions with the system (Farrell, 1999). When there is a discrepancy between the pilot's mental model and the way the system actually operates there is a cognitive mismatch (Franza & Fanjoy, 2012). This discrepancy can be caused by gaps or misconceptions in the mental model, by execution of certain modes or combinations of circumstances rarely encountered before, or by an inability to apply what they know about the system into context as actually experienced.

Nadine Sarter, human factors expert at the University of Michigan, had identified that this cognitive mismatch is the cause of one of the most common types of pilot error: mode error (Chow et al., 2014; Sarter, 2008). Mode error occurs when there is a mental disconnect occurring during the transition from the current to the actual future state of a system versus to the presumed future state of the system. An often stated, causal factor in aircraft mishaps is mode error in which the pilot action is "appropriate for the assumed, but not the actual state of automation" (National Transportation Safety Board, 2013, p. 68).

Ultimately, a lack of mode awareness may lead to a situation called an automation surprise (Rushby, 2001; Sarter et al., 1997). Under this formulation, a breakdown in mode awareness leads to errors of omission where the system commands some action that the pilot is not anticipating and for which they are unprepared. Such surprises are experienced even by those pilots considered highly experienced on automated aircraft. These problems are exacerbated when in a non-normal and time critical situation, such as an in-flight emergency or short-notice change in the route of flight (Sarter et al., 1997). Automation surprises begin with an improper assessment by the pilot caused by a gap in understanding of what the automated system is commanding, and miscommunication between crew members on what they are going to do. The potential for this problem is greatest when there is a loss of situational awareness leading to a mode error (Woods & Sarter, 2000).

A closely correlated problem with automation surprise is known as out-of-the-loop syndrome. In this situation the automated system acts to "remove the pilot from the active control loop so that the pilot loses familiarity with the key system elements and processes for which he or she is responsible" (Kantowitz & Campbell, 1996, p. 125). The effect can be to reduce the pilot skill and awareness to the point that she or he is no longer effective when faced with a system emergency, or they fail when required to jump back into the active control loop and directly control the system. Studies have shown that many pilots feel that they are often out of the loop and were concerned about taking over if the automation failed (McClumpha et al., 1991). The Air Transport Association—the major representative of U.S. based commercial airlines—has expressed its concern that some pilots appear reluctant to take control of an aircraft from an automated system (Parasuraman & Riley, 1997).

Another related problem is the aspect of human nature that people do not generally make good system monitors. "Researchers have obtained considerable evidence demonstrating that increasing automation and decreasing operator involvement in a system control reduces operator ability to maintain awareness of the system and its operating states" (Strauch, 2004, p. 224). Historically, routine tasks are automated and high workload tasks are not. This type of automation has been called clumsy and can actually increase the chances for operator error. Maintaining vigilance when workload is excessively reduced can lead to boredom and then complacency (Strauch, 2004; Young et al., 2006). The combination of boredom, delayed time from user input to system feedback, and overreliance on automation (based on experience showing a low failure rate) tends to make pilots more complacent and make it more difficult to detect and recover from errors (Sarter et al., 1997).

Another safety issue related to automation has been identified by the NTSB as pilots not having specific knowledge and proficiency to operate aircraft equipped with glass cockpit avionics (Franza & Fanjoy, 2012). In fact, a major impediment to the successful implementation of automation is the difficulty many operators have in understanding the automation, even when the system is working as designed and they are paying close attention, which speaks to poor interface design and inherent complexity (Endsley, 1996). Pilots have reported difficulty in what the boundaries of performance are, what the automation is doing, and how to intervene effectively when problems arise. Often these problems do not present themselves until the pilot has had significant experience with unusual situations that force the pilot to appraise his or her ability to understand and feel competent with automation. It appears that the rapid pace of technological advance in automation is outstripping the pilot's ability to comprehend all the implications for aircraft performance, requiring a significant increase in the amount of training provided (McClumpha & Rudisill, 2000).

For pilots, one of the most significant areas of concern is the potential loss of manual flying skills that may accompany a high level of automated flight (Geiselman, Johnson, & Buck, 2013). Though automation can assist in avoiding loss-of-control accidents, the lack of hands-on manipulation can lead to a deterioration of skills. Many operators either strongly encourage or even demand that flight crews use all automation available to them. Other operators leave the decision to the crew. Also, long-term participation as a system monitor can naturally lead to a reduction in base-line skills, mental models with a weaker representation of reality, and reduced decision making abilities. But whatever the airline policy or experience level, crews need training in when higher or lower levels of automation are appropriate and part of that training should include manual flight skill training such as that achieved in a simulator or real-world aircraft manipulation (Young et al., 2006).

Hysteresis

As more potential problems are identified related to cockpit automation due to the sharing of functions between the human and the technology, it should be recognized that operator workload varies significantly over time depending on the relative dominance in controlling of the information flow to the aircraft. Research has suggested that variation in task demand and an individual's expectations with respect to that demand may play an important role in performance (Smolensky, 1990). Extremes in demand can be exemplified by example cases: a sudden increase in the number of technology signals that must be detected and processed following a period of inactivity in which feedback was infrequent or insignificant, or a rapid reduction in the requirement for signal detection and processing following a period of sustained high workload.

Such extremes are indicative of the hysteresis effect, in which a system responds differently to identical inputs depending on the direction in which the system is being driven, or in the case of pilots, the relationship of task performance relative to the task demand (Farrell, 1999). Although hysteresis is generally used as a collective term in which work load history affects performance in future time periods, it actually applies to two different patterns of operator performance under varying workloads (Smolensky, 1990). The first is the tendency for a decrease in operator performance when the person is required to work beyond a certain work load level even after the load had decreased sharply. The second is the phenomenon of a sudden decrease in performance following a shift in visual load.

The implications of the hysteresis in the aviation environment is that overload should be avoided because an operator may develop optimal strategies that are effective for high workload conditions but when applied to low workload conditions may actually lead to the omission of critical information (Farrell, 1999). For the pilot, as the level of automation increases, the level of manual control of the aircraft decreases, but the frequency of many other activities does not change (Damos, John, & Lyall, 2005). Automation often has the effect of redistributing workload over time rather than eliminating workload for the pilot. The Delft University of Technology in the Netherlands experimented to determine the task demand relationship of various automated functions (Spencer Jr., 2000). They found there is a breakeven point where more automation actually lowers performance as predicted by the hysteresis effect.

Existing Classification Systems

Given the accepted role of human action in maintaining aircraft safety, it was natural that several researchers would look for ways of identifying and classifying that impact, though usually with a somewhat narrow focus on the researcher's particular area of interest. One of these is known as the Extended Control Model (ECOM) which describes control of a complex system such as an aircraft as taking place on several layers of control simultaneously through the use of concurrent control loops (Hollnagel, 2007). Some of these loops are closed, some are open, and some are mixed. Each layer is described in terms of a basic construct-action-event cycle and can describe interactions between different layers and among multiple participants within a layer. However, this theory does not formally define the number of layers and tends to focus more on the cognitive aspect (construct) of the players and less on the action and event aspects.

The Air Traffic Analysis and Classification System (AirTracs) is another framework used for systematically analyzing the impact of human performance on air traffic control safety events (Berry & Sawyer, 2013). It uses a set of causal factors adapted from the U.S. Department of Defense to categorize trends and look for items that may be considered causal. While this model is instructive in identifying causation for some situations, it is primarily focused on the air traffic control aspect of the system and requires a consensus on causal factors in order to arrive at conclusions, which requires a larger body of experts to participate.

One method developed by the FAA to proactively identify hazards to successful human performance in the NAS environment is the Human Error and Safety Risk Analysis (HESRA) tool (Sawyer et al., 2011). Under this methodology prospective hazard assessment starts with a task level description of the proposed activity. Then each activity is given a risk assessment according to the priority classification listed in Table 2. The resulting assessment is used to create design and research requirements to minimize the risk from the human component of the system. This provides a convenient classification system, but does not integrate all stakeholders and aspects of the environment.

Table 2

HESRA Risk Priority Number Classifications

Category	Definition/Action
Extremely Low Risk	No system or safety implications No further design or evaluation efforts required
Low Risk	No significant system or safety implications Unlikely that significant design, training, or procedural changes will be required
Moderate Risk	Potentially significant system or safety implications Possible that significant design, training, or procedural changes will be required
High Risk	Significant system or safety implications Likely that significant design, training, or procedural elements will be required
Extremely High Risk	Critical system or safety implications Significant design, training, or procedural changes are required before the system is deployed

Note: Taken from Sawyer et al, 2011, *Proceedings of the Human Factors and Ergonomics Society annual meeting*, p. 57.

A model developed in the field of human factors for systems beyond just aviation is the Software-Hardware-Environment-Liveware (SHEL) model (Spencer Jr., 2000). In the cockpit automation context, the software portion deals with the computer programs, operating procedures, and decision algorithms, while the hardware parts of this model refer to the cockpit layout and control technology being used. The liveware represents the pilot, and the environment is concerned with mental and physical stress being imposed on the human pilot. Under this conception, the pilot interacts and gets feedback from each of the other parts of the system. Because this system was designed as a generic human factors analysis, the focus is on the man-machine interface and attempts to provide insight on ideal feedback to the pilot for error mitigation.

The Step Ladder Model was developed by Rasmussen as a symbolic processing model based on human performance (Faulkner, 2003). It defines the mental activity that takes place between the initiation of a response to the feedback received in the cockpit and the actual action taken by the pilot. Rasmussen intended the sequence of actions that make up this activity to be based on rational, causal reasoning that connects various mental models of the pilot. He defined the states in the sequence as activation, observation, identification, interpretation, task definition, procedure formulation, and execution. In this model, the increasing complexity of modern aircraft technology could lead to increased divergence from an appropriate sequence due to automation design features such as unfamiliar flight controls or silent mode transitions made by the technology.

Another useful system of classification historically used to categorize causal factors in aircraft mishaps was developed by Krokos and Baker and is known as the Aviation Causal Contributors for Error Reporting Systems (ACCERS) (Hendrickson, 2009). Under this system, reports were analyzed for factors that could be considered as causal or contributory and placed into categories—originally nine and later condensed to seven as identified in Table 3.

Table 3

ACCERS Causal Categories

Initial 9 Category Solution	Revised 7 Category Solution
Procedural Issues or Deviations	Policies or Procedures
Error Made by Other People	Human Error
Pilot Error	
Weight and Balance Issues	
CRM or Physiological Factors	Human Factors
Organizational Factors	Organizational Factors
Equipment Limitations or Failures	Hardware
Weather	Weather or Environment
	Airspace or Air Traffic Control
Unexplained Events	

Note: Hendrickson, S., 2009, *The wrong Wright stuff: Mapping human error in aviation,* p. 13.

Cumulative Effect Model

Regardless of the method of identifying and measuring effects of automation

concerns, there is a strong consensus that the cognitive formulation, or mental models, of

what is happening with that technology is a primary consideration both within the cockpit

crew, and even beyond to support players such as air traffic controllers, airline executives

and dispatchers, and system designers. Rouse and Morris (1986), explain that "mental

models are the mechanisms whereby humans are able to generate descriptions of system

purpose and form explanations of system functioning and observed system states, and

predictions of future system states" (p. 7). They are important to shaping how the

stakeholders act in various situations and ensuring proper action is taken according to the

situation demands by drawing relevant inferences and understanding how other players

will act. Shared mental models represent an organized understanding or cognition among

all the members of the team and allow for the flexibility to shift knowledge across individuals in response to novel situations (Martinez, 2015). Given the roles that all the actors can take on in a typical aircraft flight—developing the equipment and standard operating procedures, planning the flight, providing tactical clearances, and executing the flight—the model that is most applicable to the shared roles and consistent mental model is Reason's Cumulative Effect Theory, or Swiss Cheese model.

Reason (1990) argued that accidents are caused by complex sociotechnical systems and organizational factors that operate within social norms of organizational policies, procedures, and goals. Human error is inevitable in such systems because of demands for success made by the imperative for financial gain, service rendering, or provision of a product recognized as valuable. His Swiss Cheese model represents the ecosystem of the aircraft operating environment and where errors can be introduced to that system. In the visualization as shown in Figure 3, there are a sequence of holes in the organizational defenses that can be attributed to both active failures and latent conditions. When these holes line up the trajectory of an accident is allowed to progress unimpeded. Although this model was seen as a good vehicle for integrating all human error perspectives into the accident chain of events, applying this theoretical model as an analytical tool was enhanced by the inclusion of the Human Factors Analysis Classification System (HFACS) developed by Wiegmann and Shappell and shown earlier in Figures 4, 5, 6, and 7 (Martinez, 2015; Shappell & Wiegmann, 2000). In examining the tradeoffs in aviation between operational efficiency and safety, such an analytical tool is ideal for considering the impact of human operators within the system (Smolensky, 1990).

Implications for FAA's NextGen Program

While the question of the impact of cockpit automation on aviation safety is already a pressing question, the next decade will be even more dependent on technology as the FAA implements a new concept for the NAS known as NextGen (Federal Aviation Agency, 2011). NextGen is described as a series of interconnected policies, systems, and programs that use the enhanced capabilities of new technology to make revolutionary changes to the way the current U.S. aviation system operates. By sharing satellite surveillance and information technology networks, both air traffic controllers and aircraft operators will be precisely aware of the location of aircraft in the air and on the ground preparing for takeoff in order to increase safety, reduce delays, enhance efficient use of the airspace, and mitigate environmental impacts of commercial aircraft operations.

A central tenet to the NextGen program is the implementation of a concept called free flight, which is defined to be "a safe and efficient flight operating capability under instrument flight rules in which the operators have the freedom to select their path and speed in real time" (Hollnagel, 2007, p. 409) instead of being directed in their flight path by air traffic control. The paradigm shift from directed flight to free flight will also shift the responsibility of safety from the ground-based controllers to a more shared responsibility with cockpit crews, and will require the use of new technologies that will fundamentally change the way humans, machines, and human-machine systems operate within the U.S. airspace (Thomas & Rantanen, 2006). This concept is quite a radical change from the status quo, such that when even experts are first confronted with the concept, the initial reaction is that it sounds like a dangerous idea (Hoekstra et al., 2001).

However the FAA plans to address the safety concerns by proactively identifying hazards and managing risks through a continuous analysis of data, while still imposing policy based air traffic restrictions (Federal Aviation Agency, 2011). Robert Wright, manager of the General Aviation and Commercial Division of the FAA, discussed the need to address safety under the free flight idiom by the development of FAA/Industry Training Standards (FITS) (Franza & Fanjoy, 2012). FITS uses scenario-based training on risk management, decision-making, situational awareness using current weather, data communication tools, and satellite positioning technology to prepare pilots for the increased responsibility (Franza & Fanjoy, 2012; Salas et al., 2010). However, from a risk management perspective, there is a pressing need to integrate human factors considerations with the implementation of NextGen technology to address the possible impacts—both positive and negative—from this new program (Berry & Sawyer, 2013; Federal Aviation Agency, 2011).

The advent of NextGen has a strong potential to exacerbate any existing problems with automation and to also create new ones. Kathy Abbott of the FAA has said "the data suggests that the highly integrated nature of current flight decks and additional add-on features have increased flight crew knowledge and introduced complexity that sometimes results in pilot confusion and errors during flight deck operation" (Chow et al., 2014, p. 67). The complexity introduced by the technology is compounded by the fact that aircraft manufacturers mix avionics from a range of suppliers into a coherent whole, but each of those suppliers designs their equipment using proprietary standards and interfaces (McClumpha & Rudisill, 2000). Many airlines have announced plans to transition to paperless cockpits in which a tablet such as an Apple iPad® or Microsoft Surface® is loaded with software versions of all the charts, logbooks, and handbooks traditionally carried as paper hardcopies (Takahashi, 2012). The FAA has recently authorized the complete replacement of paper with electronic versions known as Electronic Flight Bags (EFBs). Such a transition is largely dependent on the manufacturer to ensure consistency and safety. Although FAA airworthiness certification is necessary for aircraft and any device which is physically connected to it, hardware such as tablets that do not connect to the airplane for anything other than a source of power do not require certification (Federal Aviation Agency, 2012) and are minimally regulated under 14CFR Part 91, General Aviation Operations (Joslin, 2013).

Despite the limitations and complexity, new aircraft development remains primarily focused on technology because there seems to be a widespread belief that greater investments in automation promises lower expenditures on developing human expertise and because it provides an economic benefit by performing a task more accurately or reliably than a human operator (Parasuraman & Riley, 1997; Woods & Sarter, 2000). However, data have consistently shown that new types and levels of automation drive new knowledge and training requirements for crewmembers (Woods & Sarter, 2000). Studies have shown that teaching cockpit automation skills is necessary and viable, that classroom training can be successful in this endeavor, and that significant skill development can be accomplished without the use of the actual aircraft or simulator (Casner, 2003). This is feasible when the training is human centered and focuses on problem-driven solutions, activity-driven training, and uses context-bound scenarios (Woods & Sarter, 2000).

Future FAA Direction

Pilot deviations and unexpected aircraft performance will always exist. To achieve the next level of safety, the FAA feels it must work with industry to build on existing proactive accident prevention programs through the implementation of new tools and metrics to anticipate risks, identify and mitigate accident contributors, and manage safety resources to maximum effect (Federal Aviation Agency, 2011). Additionally, other government agencies are contributing advice to the FAA's direction. The Department of Transportation Volpe Center has been actively involved in identifying and documenting human factors issues associated with the flight procedures that use the latest satellite-based technology as a requirement for flight (Chandra & Grayhem, 2012). The NTSB has also opined that the FAA should develop more stringent standards for the surveillance of passenger carrying aircraft operators. They have determined that the rapid growth in flight operations warrant increased oversight and specifically called for: verification of adequate surveillance and evaluation staffing, increased staffing in certain areas, and augmentation of staff with airplane-specific qualified inspectors (Olson, 2000).

The FAA has also taken a more proactive stance with regard to training. They have currently endorsed the sixth generation of CRM, which focuses on Threat and Error Management (TEM) (Wagener & Ison, 2014). The new training not only looks at optimizing the person-machine interface and the acquisition of timely information, but also the interpersonal activities in the cockpit to maintain an effective team with appropriate levels of situation awareness. It also endorses preemptive strategies of threat recognition, avoidance, and management. Although CRM efforts appear to have an impact because decision based errors have been decreasing, skill based errors have not, highlighting the need for focused crew training (Jones et al., 2013). U.S. industry has been encouraged to provide training and given the freedom to develop appropriate training by the FAA, but the minimum training standards established by the regulator may not be enough to prevent aircraft accidents (Wagener & Ison, 2014). This is because crew training is a major financial burden for operators, and there has historically been a tendency to minimize expenditures on training with the intent that pilots get the rest of their knowledge through operational experience (Young et al., 2006).

The combination of airline management and economic pressures has had broad impact on the confidence of pilots in managing highly automated aircraft (Sarter & Woods, 1994). A minimum level of training should condition pilots in the general knowledge of how aircraft automation works in the context of where it is being utilized, go beyond how to operate the system to include how the system operates when not receiving pilot inputs, teach how to locate and interpret the various indications about which options are active and what target values are set, and show how to operate the system when faced with abnormal and infrequent situations inflight.

Historically, the FAA has been hesitant to dictate too much to airlines on how they run their business, preferring to establish standards and leave the question of how to achieve them to the airlines, following up with inspections to ensure compliance (Federal Aviation Agency, n.d.). With the change in paradigm brought by the NextGen program, the FAA has started to be more proactive in this area. The Performance-Based Operations Aviation Rulemaking Committee and the Commercial Aviation Safety Team have now highlighted issues in human-automation interaction by pointing out the importance of crew understanding of automation states and highlighting the importance of design factors in mitigating the complexity of the systems (Silva & Hansman, 2015). The FAA has also taken a world-leading role in introducing new ground-based technologies to assist aircrews with the new responsibilities from shared control with air traffic controllers including the development of a digital communication protocol adopted world-wide and an air-to-ground network to share information in real-time with aircraft, airline operation centers, and ground-based controllers (Amalberti, 1999).

Summation

A review of the literature to date shows that since the beginning of flight, engineers have looked for ways to incorporate new technology to ease the burden on the pilot, enhance safety of the aircraft, and improve efficiency of air traffic. Much of this technology has had a significant effect in improving flight safety. However, along with those improvements have come the negative side effects of how the human pilot deals with the technology being place on the aircraft. The scientific field of human factors research has looked at much of this impact and performed research on how to mitigate problems from the interaction of human and machine.

The rapid advance in technology in the recent past has caused cockpit automation to outpace the human factors designs needed to smooth the transition for the pilot on the flight deck. The role of the pilot has changed from one of hands-on control of the airplane to that of a supervisor and monitor, only intervening in abnormal situations. Because these are not the roles humans are naturally suited to perform, the ability of technology is now approaching the limits of the person, resulting in a different set of problems than in the beginning of aviation. New avenues are being investigated to address the ability mismatch between people and machines such as training, user interfaces, and enhanced monitoring. However, the very nature of the automation has the possibility to create issues as the workload on the pilot is rearranged from historical norms.

Various schemas have been derived to analyze the impact of automation on aircraft and pilot performance. The Swiss Cheese model (Reason, 1990; Wiegmann & Shappell, 2001) is the only one that satisfies the requirement for trend analysis across different types of mishaps, allows for a strategic view of the process beyond the discovery of pilot error, and focuses areas for intervention with new policy strategies.

As the FAA continues to implement the vision of an increasingly automated and computerized NextGen program, the weaknesses currently observed in the automation of aircraft will likely become more pronounced through increased volume. The time is therefore ripe for analysis of the current state of flight deck automation and the policy surrounding its use. Based on the findings of this analysis, new directions for the implementation of NextGen can be considered to maximize aircraft safety in the NAS.

CHAPTER III

METHODS

Cockpit automation has produced many benefits in terms of efficiency, error detection, and fuel and emissions savings, but these benefits have a cost in terms of loss of situational awareness, mode confusion, automation surprise, and deterioration of pilot manual skill. The FAA was created by the United States Government to ensure the safest and most efficient aviation system possible. They execute this mission through various tools at their disposal including Title 14 of the CFR, ACs, Airworthiness Directives, Orders, and Notices (Federal Aviation Agency, n.d.). Though there has been a great deal of study in human factors issues related to automation and about aviation safety, there has not been significant study on the impact of FAA policy directives on the use of automation and its effect on safety among passenger carrying operators. Because the general public is placing its trust in the FAA to ensure their safety, it is critical that all possible methods of improving safety be examined.

This study attempted to determine if any policy guidance has been developed or could be developed that could address identified automation concerns with the result of improving safety and efficiency of the U.S. aviation system. The FAA publishes all of its policy guidance on its website and assures compliance through various systems of inspections on the pilots, dispatchers, air traffic controllers, and airlines performed on a regular basis as prescribed in its regulations. The licensing and certification requirements for aviation make it one of the most regulated and supervised transportation modes. Should new policy proposals be accepted by the FAA, the mechanism for implementing this guidance is already in place.

Research Questions

The goal of this study was to analyze the effects of advanced technology and flight deck automation on aviation safety. This objective included the automation equipment and technology itself, the human understanding (and capability to understand) the automation, and the interface between the automated equipment and the human in the system. The null hypothesis under this construct was that automation has no significant effect on the severity of aircraft mishaps and therefore the existing guidance was satisfactory to ensure the NAS was implementing the use of flight deck technology in the safest way possible.

The foundational framework of Reason (1990), as expanded by Shappell and Wiegmann (2000) shown in Figure 3 address each of the levels of influence that the FAA governs. In this context, the organizational influences represent the system designers and manufacturers of the hardware and software as certified by the federal government or the airline standard operating procedures and guidance; the supervisory layer represents airline oversight, planning, and training requirements; the precondition layer is the tactical airline planning and pilot preparation functions performed before flight; and the acts layer would be the actual implementation of automation and the performance of the pilot on the flight deck. In order to determine whether any of these factors play a significant role and whether they can adequately be addressed through policy guidance, the following questions were addressed:

1) Does the implementation of automation technology have a significant impact on the incidence of aircraft mishaps in United States aircraft?

    a. If there is an impact, what is the impact of human factor issues?

    b. If there is an impact, what is the impact of technological deficiencies?

    c. If there is an impact, what is the impact of human-machine interfaces?

2) Is the current policy guidance on cockpit automation sufficient with respect to:

    a. The scope of guidance provided to passenger carrying aircraft?

    b. The relevance of guidance to the current state of technology?

    c. The compliance and enforcement of such guidance on airlines?

3) Does the type of training received by cockpit crew and the level of experience with cockpit automation change the relationship of flight deck automation and aircraft mishaps?

4) What are the policy actions that have shown an ability to mitigate negative effects or enhance positive effects of flight deck automation on aviation safety?

Design

The investigation of aircraft mishaps has a long history dating to the Air Commerce Act of 1926 (National Transportation Safety Board, n.d.-a). Since that time, investigators and researchers have used various techniques to analyze and model behavior that causes mishaps. Shih, Ancel, and Jones (2012) recount many of these including object-oriented Bayesian networks, fault trees, event trees, and event sequence diagrams. However, all of these methods are designed to investigate the root cause of a single given mishap. They do not attempt to look for trends across accidents. Given the goal of analyzing the impact of specific causes across accidents, another methodology needed to be used that compared similar causes across different events and different circumstances. The field of human factors analysis provided this vehicle.

The study was conducted using an explanatory sequential mixed method design using elemental methods (Saldaña, 2013) derived from the epistemological nature of the research questions. A pragmatic approach to the narratives was used where the descriptive factors were analyzed for patterns to capture the essence of the causal factors and map them to concepts that formed emergent themes. These themes naturally grouped into the four levels of the Swiss Cheese model and were used to answer the research questions that drove the study. The sequential design was due to the timing of data analysis and interpretation of the results. In accordance with the methodology described by Creswell and Plano Clark (2011), this method of design relied on an initial quantitative phase designed to address the research questions. This was followed by a second, qualitative phase that expanded on initial quantitative results to explain and understand the results. Following Creswell and Plano Clark (Creswell & Plano Clark, 2011), the study was undertaken from the perspective of pragmatism, which has wide acceptance as most applicable to mixed methods research because it focuses on the consequences of the research, on the primary importance of the research questions, and on the use of multiple methods of data collection.

With respect to this study, both quantitative and qualitative data were obtained through the publicly accessible databases published by the NTSB Aviation Accident Database and Synopses (National Transportation Safety Board, n.d.-a) and the National

Aeronautical and Space Administration (NASA) ASRS (National Aeronautics and Space Administration, n.d.). According to their web site "The National Transportation Safety Board is an independent federal agency charged by Congress with investigating every civil aviation accident the United States" (National Transportation Safety Board, n.d.-a). Accordingly, they have an exhaustive database of investigative reports on all aviation accidents and make these reports available to the public in the interest of enhancing safety by learning from past experiences. The data available from this database includes quantitative demographic data on the accident as well as qualitative data from the subjects of the investigations, and includes interpretations from qualified experts on causal factors. In order to ensure full and open disclosure in NTSB investigations, the investigatory board's analysis of factual information and any subsequent determination of probable cause is non-judicial and cannot be entered as evidence in a court of law (National Transportation Safety Board, n.d.-a).

The ASRS was established as a result of the investigation into the TWA 514 accident listed in Appendix A, in which information from a prior near-miss situation was not passed on to the crew of the accident aircraft. Its mission is to collect data submitted voluntarily from pilots, controllers, or other stakeholders with regard to aviation situations or incidents that do not rise to the level of an accident to be investigated by the NTSB (National Aeronautics and Space Administration, n.d.). The program uses this data as an educational tool for pilots to learn from the mistakes and mishaps of other pilots through various methods including a routine publication, an alerting system, and through research studies. The FAA, through Advisory Circular 00-46E, has established the regulatory authority and goals of the system (National Aeronautics and Space Administration, n.d.). Most important to the usefulness of both the validity and comprehensiveness of the data is the specification in the Circular that explains the immunity clause of the report. Under this guidance—which is also confirmed in Title 14 of the CFR (14 CFR) part 91, § 91.25—any report made into this system shields the reporter from disciplinary action or reporting to the licensing divisions of the FAA. All reporting is confidential and the website highlights the fact that there has not been a breach of confidentiality in more 34 years (National Aeronautics and Space Administration, n.d.). The result is that the aviation community has a high regard for the program and a strong incentive to self-report any incident or mishap that they experience which falls short of the NTSB mandate.

An explanatory sequential mixed methods design was used because the nature of the research questions required an understanding of which factors are shown to be impactful on the safety of flight. This question lent itself to a quantitative analysis of mishap reports to determine which factors have been shown to be either causal or contributory. Once those factors were identified however, a qualitative analysis was needed to understand what aspects of that factor were relevant and what actions or activities were significant in mitigating negative influences or highlighting positive influences. Creswell and Plano Clark (2011) explain that this design is most useful when attempting to identify and assess trends with a quantitative procedure, and to then be able to explain the mechanism or reasons for the resulting data.

Sample

Since the organization was established, the NTSB has investigated more than

132,000 aviation accidents, including accidents within the United States and those

occurring outside the United States involving United States registered aircraft (National

Transportation Safety Board, n.d.-a). They have used that information to issue over

13,000 safety recommendations, though they have no enforcement or regulatory

authority. Therefore, they take great care in maintaining a reputation for thorough,

accurate, and independent investigations. The results of those recommendations are used

to produce timely and thoroughly evaluated recommendations to the aviation community.

They also make all of their investigative reports available to the public on the NTSB

website after sanitizing the reports for personal information. The reports of the NTSB are

limited to those accidents which they investigate. The threshold they use to classify an

incident as worthy of investigation is:

> An accident is defined as an occurrence associated with the operation of an
> aircraft that takes place between the time any person boards the aircraft with the
> intention of flight and all such persons have disembarked, and in which any
> person suffers death or serious injury, or in which the aircraft receives substantial
> damage. An incident is an occurrence other than an accident that affects or could
> affect the safety of operations. (National Transportation Safety Board, n.d.-a)

The ASRS is designed to capture all mishaps and incidents that fall short of the

NTSB threshold for investigation. According to the ASRS website, any stakeholder, to

include pilots, controllers, flight attendants, maintenance personnel, dispatchers, or any

other person with interest in the NAS, can report actual or potential discrepancies

involving the safety of aviation operations (National Aeronautics and Space

Administration, n.d.). This is a broad definition that can cover not only mishaps in flight, but also air traffic control, maintenance, ground-based movement, or crew-to-ground communications. Between the broad scope of concerns, the promise of immunity, and the demonstrated value of the program over the years, the reporting system now receives more than 7,500 reports a week and has processed over 1.3 million reports since its inception in May 1975. Although the system was established by the FAA, from the beginning it was recognized that the promise of immunity and objectivity was critical and would be enhanced by the involvement of a third party as the administrator. Therefore, NASA took on the responsibility. A report can be filed online or via paper submission, but the format is similar for either and is shown in Appendices C and D.

Between these two sources, a majority of aircraft mishaps and accidents were available for review. In each of these databases there exists demographic information on the type and location of the incident. Furthermore, both datasets provide narratives on the incident from the point of view of the pilot (when available), witnesses, and—in the case of NTSB reports—expert investigators providing their considered opinions. Although the potential population for analysis using both of these datasets was in the hundreds of thousands, the sample was significantly limited by restricting the search to examples that were identified as having either a primary or ancillary cause related to the automation of cockpit activities or the human interaction with such automation. Furthermore, because the issue was first recognized and addressed following the American Airlines flight 965 crash in Cali, Colombia in 1995 as shown in Appendix A, searches were limited to the timeframe after that investigative report was published in 1996, which resulted in a total number of cases equal to 174,094. Because all data is currently available on public websites, no IRB approval was required to access the data.

For quantitative analysis the outcome of interest is not continuous, rather it is ordinal. The severity of the mishaps determine whether or not the NTSB performs an investigation (National Transportation Safety Board, n.d.-c). They classify those mishaps on a scale of minor, serious, or fatal. All ASRS reports are for mishaps in which there were no reported injuries, otherwise the NTSB would be required to investigate. Because the objective was to improve safety of passenger travel and the data came from both NTSB and ASRS, the outcome of the mishap was classified as no injury, minor injury, serious injury, or fatality. These outcomes can be naturally rank ordered, but cannot be placed on a numeric scale that has real-world meaning for the difference in categories. Therefore, the outcome was considered ordinal. The predictor variables were dichotomous because the automation equipment in question either did or did not exist in the causal finding being reported. The model that allows for the dependent variable to be ordinal, while having dichotomous independent variables is an ordinal regression. The ordinal regression is an extension of the general linear regression using a PLUM (polytonomous universal model, where polytonomous indicates more than two distinct categories) and is available in SPSS (Norušis, 2012), therefore that was selected for this analysis. However, because more than one factor can be identified as either causal or contributory in an investigation, any given individual incident could show up more than once in a search. To maintain independence, each mishap was limited to one representation in the analysis as determined by the primary factor listed in the causal narrative. Following the identification of trend issues by quantitative review, the elemental coding method of qualitative analysis described in the qualitative analysis section below was accomplished using a review of the narratives provided in the investigative reports to look for consistent themes and areas of commonality that may be addressed for future safety considerations.

Variables and Measures

According to the Swiss Cheese model, various automation and human-automation interactions could impact all four levels. The factors included in the quantitative analysis are listed in Table 4. Although there are hundreds of automated functions on modern aircraft, only the systems which require input or feedback from the pilot and which can take control of the aircraft are selected to be independent variables. The combined database of NTSB and ASRS incidents was filtered to limit the results to mishaps occurring after 1996 and which resulted in an actual mishap vice a near-miss, resulting in a total case count of 65,837. Although there was variation in the attributed causes due to pilot self-report differences and different investigative teams making up the panel for each NTSB report, a level of standardization was imposed by the multiple choice nature of the ASRS and the level of training NTSB panel members must receive before participation, as well as post-report validation procedures by NASA and the NTSB. All data for this and the follow-on qualitative analysis was obtained from open-source data published on the public websites.

Table 4

Model Variables and Count in Dataset

Variable	Identifier	Description	Count
Mishap Severity			
	Minor	Any injury which does not achieve the definition of severe or fatal	65,033
	Serious	Any injury which requires hospitalization for more than 48 hours, commencing within seven days from the date of the injury. Also includes other injuries, including fractures of any bones (except simple fractures of the toes, fingers, or nose), severe bleeding, internal organ injuries, and severe burns.	596
	Fatal	Any injury which results in death within 30 days of the accident.	208
Automation Technology			
	Database	Any technology which accessed the internal aircraft database of flight parameters, navigation waypoints, or terrain and obstacles	1,835
	Automated Controls	Any piece of automation which is allowed to take over control of some function from the pilot	111
	Pilot Interface	Any hardware or software that allows transfer of information from the pilot to the aircraft or the aircraft to the pilot	349
	Digital Communications	Any communication equipment that is provided digitally including voice or data whether or not pilot intervention is required	259
	Flight Management System	Any piece of equipment that provides for the transfer of preplanned flight information to the navigation or display systems of the aircraft	898
	Flight Computer	Any computer installed on the aircraft that performs some function automatically without required calculation from the pilot	795

Note: Collated from the NTSB website, n.d.,
https://www.ntsb.gov/_layouts/ntsb.aviation/index.aspx.

Quantitative Analysis

Research question number one required a quantitative analysis. The question was:

1) Does the implementation of automation technology have a significant impact on the incidence of aircraft mishaps in United States aircraft?

    a. If there is an impact, what is the impact of human factor issues?

    b. If there is an impact, what is the impact of technological deficiencies?

    c. If there is an impact, what is the impact of human-machine interfaces?

An ordinal regression is developed from the premise that there is an underlying, latent, and unknowable continuous variable of outcomes that represents itself as the discrete outcomes observed. In this case the outcomes are severity of mishap. Based on the latent value of the underlying variable the value of a given ordinal outcome is observed. The value of this variable is known as the threshold (Fahrmeir, Kneib, Lang, & Marx, 2013). Because of this methodology, a threshold valued is calculated, but an intercept is not modeled in the equation. An intercept value would only move the scale of the outcome observed, which is not of concern in the model. The resulting generic ordinal model is:

$$ln\left(\frac{\rho_j}{1-\rho_j}\right) = \alpha_j - \sum_{i=1}^{k} \beta_i x_i$$

Where:
$$\begin{cases} j = \text{number of categories} \\ \alpha = \text{threshold} \\ k = \text{number of predictors.} \end{cases}$$

From this model, it can be observed that although the threshold value changes for each outcome of interest, the coefficients of each predictor variable are constant across all outcomes (Fahrmeir et al., 2013; Norušis, 2012). Therefore, all predictor coefficients have the benefit of being constant regardless of outcome. Another benefit of this model for ordinal data is that the spacing between the variables is not significant and we need not be able to specify the magnitude of difference between a fatal and serious mishap. The model instead measures the odds of observing the outcome relative to all outcomes and that ratio remains proportional over the domain of outcomes (Fahrmeir et al., 2013).

To answer this question, an ordinal regression was run and the analysis was used to determine which causal factor—as a function of what type of technology, human factors, or human-machine interface—had a significantly higher impact on the safety of flight in terms of mishaps reported to either NTSB or NASA. (Norušis, 2012). Because ordinal variables do not naturally form linear functions, a link function is required to map the probability of each outcome to a linear model. The link function is so named because it links the left side of the equation with the independently determined predictors on the right side of the equation. There are five potential link functions commonly used to make this connection and they tend to work well in certain scenarios. The functions are logit for when categories are evenly distributed, probit for when variables are explicitly normally distributed, Cauchit for when outcomes have many extreme values, complementary log-log when higher categories are more probable, and negative log-log for when lower categories are more probable (Norušis, 2012). The outcomes in NTSB reports were more likely to have minor injuries, so the negative log-log was tested for suitability and shown to be applicable.

This model is:

$$-\ln(-ln(\gamma)) = \beta_0 + \beta_1 x_1 + \beta_2 x_2 + \beta_3 x_3 + \beta_4 x_4 + \beta_5 x_5 + \beta_6 x_6$$

Where

β_0 = threshold value for each severity of mishap
1 = the database contributed to the cause
2 = the automated controls contributed to the cause
3 = the pilot interface contributed to the cause
4 = the digital communications contributed to the cause
5 = the flight management system contributed to the cause
6 = the flight computer contributed to the cause
Where 1…6 are binary variables

The null hypothesis was that the factors had no significant impact, and each cause or level of experience that exhibits significance at the $\rho = 0.05$ level was considered to reject the null hypothesis and to be significant. Factors were deemed either significant or not significant and if significant, the factor was examined in the sequential qualitative analysis within the context of the theoretical framework.

As with any statistical model, certain assumptions must be fulfilled for the analysis to be valid. In the case of ordinal regression several were checked prior to running the regression (Laerd, n.d.; Norušis, 2012). In addition to checking the form of the negative log-log link function above, four assumptions were checked. First was the requirement for the dependent variable to be ordinal. Second was for the independent variables to be either ordinal or dichotomous. Third was for there to be no multicollinearity. Fourth was for there to be proportional odds, meaning that each independent variable has the same effect on the ordinal variables. This is required because unlike OLS, in ordinal regression each threshold value of the dependent variable has its own equation with a different intercept, but common predictor coefficients

(Norušis, 2012). The results of these checks are shown in the results section of this study.

Qualitative Analysis

Research questions two through four could not be measured via objective means

or quantitative measures, so a qualitative analysis was necessary. These questions were:

2) Is the current policy guidance on cockpit automation sufficient with respect to:

    a. The scope of guidance provided to passenger carrying aircraft?

    b. The relevance of guidance to the current state of technology?

    c. The compliance and enforcement of such guidance on airlines?

3) Does the type of training received by cockpit crew and the level of experience

with cockpit automation change the relationship of flight deck automation and

aircraft mishaps?

4) What are the policy actions that have shown an ability to mitigate negative effects

or enhance positive effects of flight deck automation on aviation safety?

To answer these questions, the results of the quantitative analysis were used as a

roadmap for a qualitative analysis. Following the determination of significance, an

analysis of the narratives of a subset of reports was undertaken to discover underlying

causal issues and actions taken to mitigate them. The use of the qualitative research

software NVIVO was used to code, map, and identify underlying relationships and seek

out trends according to the cyclical, deductive approach presented by Saldaña (2013). As

a starting point for analysis, an established taxonomy for mishap causal factors relating to

automation that had been established by Jones et al. (2013) was used to categorize the type of factor as shown in Table 5 along with the list of most common words identified

by NVIVO in Figure 8 and the terms identified during the pre-coding phase. The

taxonomy was shown to be especially useful because it maps specific causal factor codes

used by the NTSB to more general human factors analysis categories developed by

Wiegmann and Shappell (2001) shown earlier in Figures 4, 5, 6, and 7. This initial

taxonomy was expanded to include actual data as identified from the most common word

list and by the coding procedure using structural, descriptive, and in-vivo coding methods

from Saldaña (2013). This type of coding also mitigated potential bias by correlating

codes to the actual text of the narrative rather than researcher interpretation.

Figure 8. Word cloud of 75 most commonly appearing words in NTSB narratives

Table 5

Taxonomy of Factors from Narratives

Variable		Description
Automation Centered Issues		
	Flight Crew	Development process
	Interface	Pilot/automation interface
		Human-centered design
	General Design	Functionality
		Complexity
		Levels of automation
		Automation failure
	Philosophy	Standardization
		Pilot/automation responsibility and authority
		Pilot role
Operations		
	Crew	Crew coordination
	Performance	Performance
		Use of automation
	Awareness	Situation/energy awareness
		Automation awareness
	Workload	Attention
		Workload
Pilot Centered Issues		Skill
		Confidence/trust
		Understanding of automation
		Crew personnel factors
Human Factor Issues		Communication breakdown
		Confusion
		Distraction
		Fatigue
		Human-machine interface
		Physiological
		Situational awareness
		Time pressure
		Training/qualification
		Troubleshooting
Organization Centered Issues		Company automation philosophies, policies, and procedures
		Pilot selection, training, and evaluation

Note: Taken from Jones et al, 2013, *Identification of crew-systems interactions and decision related trends*, p. 28.

The elemental methods of coding (Saldaña, 2013) were used to analyze the narratives independently of any preconceived structure. Structural, descriptive, and in-vivo coding was used to determine the concepts contained in each narrative, which were then organized into consistent themes following the patterns that appeared from the review. After performing the coding methodology, the themes were viewed under the theoretical framework of the Swiss Cheese model in Figure 3. After that, each causal factor was examined for its fitness to be assigned to one of the levels of failure: organizational, supervision, preconditions, and acts. According to this model, each level should have some responsibility for the outcome which will show as at least one significant causal factor assigned to each level. In analyzing the themes, they all fit appropriately into one or more levels of the Swiss Cheese model. The independent variables identified in the quantitative model all represented measurable issues that can be seen in one or more of the Swiss Cheese levels and which had the potential to cause or contribute to aircraft mishaps related to automation.

To perform this analysis, the data was reduced for functionality and for the most parsimonious representation. Initially, all ASRS data was removed and only NTSB data was analyzed because the NTSB has several advantages (National Transportation Safety Board, n.d.-b). First it was more comprehensive in that a minimal standard of what information gets reported is published by the NTSB and enforced through the agency's own internal validation methods, whereas ASRS data is dependent on the individual pilot to meet the published standards and there is little recourse for NASA to go back after the fact and collect missing information. Second, the NTSB investigating panel is more objective than the self-reported narratives in the ASRS. Finally, the NTSB panel is chartered to have experts in the area being investigated, which allows for a better understanding of the underlying issues as well as the ability to connect individual cases with any larger issues that an expert may be aware of, while an individual pilot may not.

The second filter was the use of only cases in which there was mishap on a passenger carrying flight. Since the goal of the study is to improve aviation safety, mishaps on non-passenger missions have limited relevance and are covered by an entirely different set of federal regulations (Federal Aviation Agency, n.d.). Furthermore, aircraft without passengers tend not to have the same technology as passenger carrying aircraft due to the high cost of the equipment not being justified without passengers. Finally, the cases analyzed were restricted to post-2004 because that is the date where the FAA first implemented a requirement for crew resource management training (Federal Aviation Agency, 2004). Before that date, comparisons of crew actions are of limited equivalence.

These initial filters reduced the volume of reports from 65,837 to 3,843. After this the individual mishap causal narratives were scanned electronically using excel to extract cases that contained words or phrases identified in the quantitative analysis and the taxonomy from Table 5. The keywords and phrases were expanded to include synonyms and wording variations to capture all the applicable mishap summaries relative to the research questions. The result was a final list of 1,021 mishap findings that fit the criteria.

Qualitative analysis was performed using the remaining accident detail narratives from the NTSB, for which the format is shown in Appendix E. Although each investigation is performed by a different group of individuals based on the timing of the mishap, the expertise that is needed for the unique circumstances of the situation, the availability of personnel, and the size of the investigatory team assembled by NTSB leadership, there is a core "go team" on call, all of whom are specialists in the standardized investigative and reporting procedures published on the website (National Transportation Safety Board, n.d.-b). This team is led by a senior Investigator-in-Charge who is required to have extensive experience with NTSB investigations, which helps ensure a level of standardization. Following the submission of a final report from the investigation, the report is sent to other experts for review and critique prior to publication. The resulting publication is a final report that has been subject to scrutiny of the original group of investigators, third-parties with direct knowledge of the issues, and other NTSB experts from the same fields of study who have not been part of the investigation to this point (National Transportation Safety Board, n.d.-b). Both the accident details in Appendix E and the full report are published to the web site. For this research, the relevant details as to the cause of the mishap are in the accident detail summary. If the findings were unclear or warranted a more in-depth analysis, the full report was consulted.

From this point, a first cycle precoding review of the data was performed. Had this data been collected by the researcher directly, an initial impression of what thoughts and concepts were significant would have been formed during the interview process. From that point a first cycle coding would have a solid basis to use as a starting point. As Saldaña (2013) referenced, most researchers perform coding both during and after data collection. Because the researcher did not collect this data, a precoding review was performed to replace the portion of the coding process that would normally have occurred during data collection. All of the remaining cases were imported into NVIVO software to expedite precoding. At that point a word frequency query was performed. With the results of that query and the initial taxonomy in Table 5 as a guide, a sampling of narratives were reviewed to identify "words and short phrases…by circling, highlighting, bolding, underlining, or coloring rich or significant participant quotes or passages" (Saldaña, 2013, p. 19) and to get an overview of the concepts that were likely to form the basis of the first cycle.

Once an understanding was obtained of the type of information in the narratives, a second cycle of coding followed the elemental cyclical coding method of Saldaña (2013) to extract concepts and themes for analysis. The elemental method of coding uses basic but focused filters to review the data and build a foundation. The coding tools of the elemental method include structural, descriptive, and in vivo. In structural coding, a phrase containing either a concept or content relative to the topic is applied to categories that form the foundation for third cycle coding. Descriptive coding takes the basic data in the narrative and assigns labels to provide an inventory of the topics discussed. In vivo coding uses the words of the source directly to provide the codes. This first pass generated the list of codes provided in the results section of this study.

Following the second cycle of coding, a third cycle was performed using pattern coding methodology (Saldaña, 2013). Pattern codes are:

> Explanatory or inferential codes, ones that identify an emergent theme, configuration, or explanation. They pull together a lot of material into a more meaningful and parsimonious unit of analysis. They are a sort of meta-code…Pattern Coding is a way of grouping those summaries into a smaller number of sets, themes, or constructs. (Miles & Huberman, 1994, p. 69)

For this third cycle of coding, NVIVO was employed to perform queries and assign nodes which collected similarly coded passages from the narratives to assess the common concepts and bring together patterns and themes. These themes formed the basis of the analysis to answer the research questions.

Verification and Validation

Methodological rigor was applied to assure verification, and validation with the intent of ensuring findings are dependable, credible, transferrable, trustworthy, and confirmable (Cutcliffe & McKenna, 1999). Techniques used were derived from Creswell (2003). The first method of verification was by referencing the literature search to identify and reconcile areas where discrepancies existed. Another method recommended by Creswell was to capitalize on the researcher's background, bracketing the researcher's experiences as both a pilot and graduate-level instructor to other experienced pilots. The 30-year experience of engagement and persistent observation in the field allowed the researcher to understand the culture, highlight information that seems incongruous or may have been misinterpreted by third parties, and focus in on portions of the narrative that are more likely to be salient to the research questions. Additionally, comparison to expert notes, and reference checking with pilot self-reports in the ASRS was performed. Finally, verification of data was enabled by ensuring transparent access to the public data and providing quoted references in the analysis process.

Validation was also accomplished using methods from Creswell (2003). One of the most illuminating was by analyzing negative cases. In the ASRS, many cases are cited in which factors identified in the analysis occurred but did not result in an injury. Review of the differences in these cases versus the NTSB cases allowed validation of the themes. Additional validation methods included expert interpretations from NTSB specialists, selecting a sufficient number of cases for analysis to achieve saturation through multiple cases, ensuring complete thoroughness of the method, and ensuring the coherence of results (Sousa, 2014). Finally, results were cross-checked with ASRS reports not used in the qualitative analysis to confirm validity of the findings.

Limitations

Several limitations were apparent in this study. First was the self-report nature of the ASRS reports. The FAA made a strong effort to incentivize the program by passing off control to NASA as a third party and immunizing reporters from penalty if they participate. They also attempted to standardize reporting by making the submission form use multiple choice options wherever appropriate. These efforts seem to have beneficial effects as shown by the receipt of over 1.3 million reports in the 40 years the program has been in existence (National Aeronautics and Space Administration, n.d.). However, even with these accommodations the final narrative on the cause of mishaps is determined by the author, who is often the same individual who made whatever mistake is being reported. Although there is a strong incentive to participate openly, there is also the human nature-based desire to paint one's self in the best possible light. It is also unknowable how many mishaps which should have been reported were not actually reported to the system.

Another limitation was that the NTSB data was collected by third parties without the opportunity for follow-up clarification by the researcher. Although the investigators were all selected for their expertise, they were not generally the same investigators across mishaps. Therefore, despite NTSB investigator training requirements, there are questions about inter-rater reliability and inability to pursue questions related to this study in depth if the investigator was not considering the impact of automation when performing interviews. The cost and time to re-interview the affected witnesses would be prohibitive for almost any researcher where such interviews would even be possible due to witness availability.

Finally, the investigators for the NTSB were not considering the theoretical framework when attributing causes. The Swiss Cheese model predicted that errors must occur at each level in order for a sequence of events that result in a mishap. It is possible, however, that the reporting and analysis methodology does not identify the contributory effort at each level. For example, the NTSB experts may have seen a sequence that results in an accident, but determined that an error at one particular level was more significant and attributed that factor as primary while dismissing the others as secondary. Although such a distinction could be significant for future studies, as a first step in determining the effect of policy, the primary factor was sufficient to analyze which actions can be addressed by the FAA or airline operators.

CHAPTER IV

RESULTS

The explanatory sequential mixed methods design of Creswell and Plano Clark (2011) was used to perform the analysis. The first research question dictated a quantitative analysis to determine the factors that were significant in the impact on aviation safety. The results from the initial ordinal regression analysis were then used to focus the issues to be analyzed for the follow-on research questions qualitatively. This analysis was directed at answering the research questions:

1) Does the implementation of automation technology have a significant impact on the incidence of aircraft mishaps in United States Aircraft?

    a. If there is an impact, what is the impact of human factor issues?

    b. If there is an impact, what is the impact of technological deficiencies?

    c. If there is an impact, what is the impact of human-machine interfaces?

2) Is the current policy guidance on cockpit automation sufficient with respect to:

    a. The scope of guidance provided to passenger carrying aircraft?

    b. The relevance of guidance to the current state of technology?

    c. The compliance and enforcement of such guidance on airlines?

3) Does the type of training received by cockpit crew and the level of experience with cockpit automation change the relationship of flight deck automation and aircraft mishaps?

4) What are the policy actions that have shown an ability to mitigate negative effects or enhance positive effects of flight deck automation on aviation safety

Performing the quantitative and qualitative analysis sequentially allowed for the research questions to be addressed. The completion of these analyses produced significant results.

Automation Impact on Aircraft Mishaps

The dependent variable of mishap severity was ordinal in that the relative severity could be rank ordered but not quantified in a meaningful way such that the difference between the three categories—minor, serious, fatal—had numerical significance. Therefore the use of ordinal regression was dictated. The SPSS program captures the use of ordinal regression to model the ordinal dependent variable with the binary nature of the independent variables where the variable was either present or not present in the mishap cause. The model used for the data in this study is:

$$-\ln(-ln(\gamma)) = \beta_0 + \beta_1 x_1 + \beta_2 x_2 + \beta_3 x_3 + \beta_4 x_4 + \beta_5 x_5 + \beta_6 x_6$$

The results of the ordinal regression are shown in Table 6 and are based on 65,837 mishap reports from both NTSB and ASRS. The nature of ordinal regression specifies a different threshold value for each severity, but the same coefficient values of β for all severities (Norušis, 2012). Because of this, the significance (ρ value) for the threshold values listed in the table being greater than 0.05 creates no concern, as the relative distance between minor, serious, and fatal mishap is not part of the study objective. For the location of independent variables, all are seen to be significant at the $\rho = 0.05$ level except for the FMS, and may reject the null hypothesis that they have no effect on mishap severity. One potential reason for this particular variable's lack of significance is that the functions and attributes of the FMS are attributed to other categories such as flight computer or pilot interface. Although the FMS is a self-contained unit, it does provide a bridge for pilot interface to the flight computer. The functions that are independent to the FMS itself could either truly be non-significant or could be attributed by investigators to one of the other systems with which it interacts.

Table 6

Ordinal Regression Model

		Parameter Estimate	Standard Error	Wald Coefficient	Significance
Threshold	Minor	31.76	23.51	1.82	0.18
	Serious	32.20	23.51	1.88	0.17
	Fatal	32.72	23.51	1.94	0.16
Location	Database	1.54	0.15	98.70	0.00
	Automated Controls	2.02	0.35	32.64	0.00
	Pilot Interface	0.57	0.15	15.13	0.00
	Digital	1.33	0.23	33.50	0.00
	Flight Management	10.22	23.51	0.19	0.66
	Flight Computer	12.12	0.26	55.19	0.00
Nagelkerke Pseudo R^2		0.02			

The results of the regression provide parameter estimates for the effect of each independent variable. Because the ordinal regression is a proportional odds model, the values cannot be viewed as direct impacts. Rather the estimated impact of the variable is calculated as exp (β) and represents the change in the odds of seeing the outcome observed. In this case, five predictor variables were calculated as significant. For the database, the odds of changing the outcome to a more serious one were 4.7 times as large, for automated controls the odds were 7.5 times as large, for pilot interface 1.7 times larger, for digital communications 3.8 times, and use of a flight computer made the higher outcome more than 183,505 times as likely. It is important to point out that these likelihoods are based on an extremely small starting baseline.

All interactions between the predictor variables were also checked for significance as part of the regression, including 2-way, 3-way, 4-way, 5-way, and 6-way interactions. None of these interactions proved significant at the $\rho = 0.05$ level. The results are not shown in Table 6 because of the large number of interactions and the lack of any significance to show.

These results clearly show that the use of a flight computer is the largest individual contributor to severity of the mishap. This is a reasonable outcome because all modern avionics and automated flight controls interact with the flight computer at some point. Although the impact may be latent to the pilot, behind the scenes all the automation requires data from somewhere—when not provided directly by the crew— and that data is processed by and received from the computer.

The Nagelkerke pseudo R^2 is 0.02. The interpretation of this is somewhat different than the standard R^2 associated with OLS regression. In this case it represents a ratio of likelihoods and can be interpreted as the improvement of the full model over the intercept model. The absolute number is relatively low which indicates that the model itself provides low predictive value for the type of outcome. However, the more relevant outcome of the model is that it implies that in cases where automation is contributory to the severity of mishaps, then the predictors indicated as having a significant ρ value are in fact contributory to the outcome.

To confirm the applicability of the model, several requirements must be met and were tested before analysis. For ordinal regression, several assumptions required of OLS regression models are not required. The requirement for normality and linearity is not needed, instead the assumptions required by ordinal regression are 1) the dependent variable must be ordinal; 2) the independent variables must be either continuous or dichotomous; 3) there must be no multicollinearity; and 4) there must be proportional odds, meaning that each independent variable must have an identical effect on the dependent variable (Laerd, n.d.; Norušis, 2012). It should also be noted that the nature of the source prevented any missing data. The database was only populated when data existed. Any missing data issues were resolved by NASA in the case of ASRS data or the NTSB prior to entering the data.

The first two assumptions are met by the nature of the data. The dependent variable is ordinal with categories of minor injury, serious injury, or fatality. The independent variables are all dichotomous, measured as either listed as a causal factor or not listed. The existence of multicollinearity is tested by running an OLS because the collinearity test only applies to predictor variables and is agnostic to the dependent variable (International Business Machines, n.d.; Norušis, 2012). The results of the collinearity test are listed in Table 7 and show all predictor variables have a variance inflation factor (VIF) below 10. This measure provides and index of how much of the variation in each coefficient is attributable to collinearity. Therefore it may be concluded there is no significant evidence of multicollinearity among the independent variables.

Table 7

Test for Multicollinearity

Model	Tolerance	VIF
Database	1.0	1.0
Automated Controls	1.0	1.0
Pilot Interface	1.0	1.0
Digital Communications	1.0	1.0
Flight Management System	0.2	4.1
Flight Computer	0.2	4.1

The last assumption to check was that of proportional odds. This was checked in

SPSS via the test of parallel lines (International Business Machines, n.d.; Norušis, 2012).

If each individual predictor has an identical effect on the dependent variable, they should

result in a set of parallel lines for each outcome variable (Norušis, 2012). The results of

this test are shown in Table 8 and show that $\rho > 0.05$ and therefore the null hypothesis

cannot be rejected that the lines are parallel and the assumption is met.

Table 8

Test of Parallel Lines

Model	-2 Log Likelihood	Chi-Square	df	Significance
Null Hypothesis	110.53			
General	87.43	23.10	14	0.06

Finally the value of the model itself was tested. As shown in Table 9, for the

measure of model fit, the $\rho < 0.05$ and therefore the null hypothesis is rejected that the

model without predictors is as good as the model with the predictors. Separate from the

measure of model parameters adding value is the question of whether the model predicts

outcomes well in the goodness of fit listed in Table 10. If the observed and expected

values from the model are close, then the observed significance level is large. In this case, that is what was observed and therefore we fail to reject the hypothesis that the model is a good fit for the data (Norušis, 2012).

Table 9

Model Parameter Fitting

Model	-2 Log Likelihood	Chi-Square	df	Significance
Intercept Only	1233.8			
Final	110.5	1123.3	7	0.00

Table 10

Goodness of Fit

	Chi-Square	df	Significance
Pearson	23.5	29	0.75
Deviance	23.5	29	0.75

Mishap Causal Findings

Following the guidelines of Creswell and Plano Clark (2011) the results of the quantitative analysis were used to focus the qualitative analysis on the impact of automation policy and training in automation on aviation safety as listed in the later research questions. The impact of the factors found significant in the quantitative analysis were used as the focal point in a qualitative review of the data as it applied to the four major failure levels: organizational influences, unsafe supervision, preconditions for unsafe acts, and unsafe acts (Reason, 1990).

From this point, each mishap causal analysis summary was reviewed and coded according to the cyclical method explained by Saldaña (2013). Initially a first-pass coding was executed using concepts from the theoretical model and the preliminary quantitative analysis using structural, descriptive, and in-vivo codes. As the summaries were reviewed, additional descriptive terms were added that identified links between data collected and expanded meaning relative to the research questions to result in repetitive patterns. The terms identified in this coding cycle are listed in the descriptive first cycle codes column of Table 11. As shown in this table the first and second cycle of coding resulted in multiple descriptions and codes that were then organized into twelve final codes that were then grouped into four themes using the guidance of the theoretical framework.

Table 11

Coding Table Derived from Taxonomy/Actual Narratives

Theme	Final Codes	Descriptive First Cycle Codes
Organizational Influences		
	Company Policies	Standardization, complexity, crew personnel factors, Company automation philosophies, policies, and procedures, Pilot selection, training, evaluation, budget, resources Operations manual, hiring, guidance, culture, climate
	Governmental Oversight	FAA, Evaluation and training requirements, inspections, Pilot responsibility and authority, compliance, regulatory, Technical orders, circulars
Unsafe Supervision		
	Training	Developmental process, qualification, continuation training
	Experience	Background, use of automation, crew composition
	Supervision	Dispatch process, crew qualification, failure to correct, Enforcement, documentation
Preconditions for Unsafe Acts		
	Human Factors	Distraction, mode confusion, situational awareness, Physiological, time pressure, human-machine interface, General design
	Crew Resource Management	Crew coordination, communication and breakdown, Confidence, trust
	Fatigue	Fatigue, crew duty day, circadian cycle, readiness
Unsafe Acts		
	Automated Controls	Technology, pilot/automation interface, performance, Troubleshooting, human-centered design, functionality, Automation awareness, automation failure, interface, Computer, FMS, autopilot, autothrottle, autobrakes, Navigation database, EFB, CPDLC, ACARS
	Distraction	Attention, workload, stress
	Overreliance	Skill, level of automation, use of automation
	Slow/Improper Responses	Understanding of automation, pilot role, Automation guidance, crew performance, disorientation, Confusion, complexity

Following this review, a second cycle of coding was performed on the descriptive codes for pattern analysis, and a third cycle coding was performed to expand and reconfigure the original codes into major themes. As Saldaña put it, the purpose of this cycle is to "capture a datum's primary content and essence." (p. 4) The emphasis was to look at patterns of similarity, frequency, or causation, then to cluster these patterns into concepts. The resulting codes are listed in the final codes column of Table 11. Upon analysis of the codes and themes identified in the cyclical analysis, it became clear that certain concepts fit clearly within the layers of failure identified by Reason (1990) and Shappell and Wiegmann (2000) and the coding table listed in Table 11 shows the categorization of those themes.

Organizational Influences

In analyzing the final list of mishap findings that related to the final three research questions, the twelve codes listed in Table 11 were distilled from over 100 initial types of factors identified in the first pass. These factors and the types of causal descriptions listed in the mishap narratives were then applied to the research questions under the rubric of the theoretical framework in order to draw conclusions on the types of policy guidance that may be successful. The first of these levels is organizational influences and several areas from Table 11 are applicable.

Company policies. Of the 1,021 cases analyzed, 102 cases listed some type of company policy as causal. Organizational influences can occur at several levels in the national aviation system. The first occurrence has the most direct control over flight crews and comes from the company operating the aircraft. Each airline has internal rules, corporate standards, flight operating manuals, and dispatch procedures that direct their crews (employees) how to behave as well as having an organizational culture that subconsciously directs employees in the absence of written guidance. Multiple instances of corporate influence were identified in the mishaps. These can be attributed to either crews failing to follow company standard operating procedures (National Transportation Safety Board, n.d.-a, case 20040513X00667) or failure of the company to provide adequate guidance. While crews following procedure may be difficult for companies to control, they have an obligation to provide sufficient guidance in the first place. Numerous examples were cited by investigators where this did not occur:

> The operator's failure to ensure that the flight crews adhered to company policies and FAA and DoD federal safety regulations, and the lack of in-country oversight by the FAA and the DoD of the operator. Contributing to the death of one of the passengers was the operator's lack of flight-locating procedures and its failure to adequately mitigate the limited communications capability at remote sites. (National Transportation Safety Board, n.d.-a, case 20041213X01969, Probable Cause Section, para. 1)

In other situations, the company operating the aircraft completely failed in their role to provide the top level guidance for the airline:

> Failure to provide its pilots with clear and consistent guidance and training regarding company policies and procedures related to arrival landing distance calculations; 2) programming and design of its onboard performance computer, which did not present inherent assumptions in the program critical to pilot decision making; 3) plan to implement new autobrake procedures without a familiarization period; and 4) failure to include a margin of safety in the arrival assessment to account for operational uncertainties. (National Transportation Safety Board, n.d.-a, case 20051213X01964, Probable Cause Section, para. 1)

Governmental oversight. There were 128 cases which listed some type of governmental oversight as causal. The government is more definitive to standardizing activities across airlines, yet it is another step removed from the activities of the flight crew. The role of government is also broader. Not only are they tasked with airline and aircrew oversight (National Transportation Safety Board, n.d.-a, case 20080925X01531), but also from the beginning of a new technological design in either the aircraft or the avionics, the government has a certification process to allow the hardware and software to operate with passengers. If certification requirements are inadequate, they can lead to mishaps in themselves. Examples include the case where the FAA failed to establish adequate certification requirements for flight into icing conditions given the limitations of the stall warning system (National Transportation Safety Board, n.d.-a, case 20050304X00267) or oversight of the manufacturer's developing specific aircraft limitations such as:

> Gulfstream's failure to properly develop and validate takeoff speeds for the flight tests and recognize and correct the takeoff safety speed (V2) error during previous G650 flight tests; the G650 flight test team's persistent and increasingly aggressive attempts to achieve V2 speeds that were erroneously low; and Gulfstream's inadequate investigation of previous G650 uncommanded roll events, which indicated that the company's estimated stall angle of attack while the airplane was in ground effect was too high. (National Transportation Safety Board, n.d.-a, case 20110403X03645, Probable Cause Section, para. 1)

Unsafe Supervision

The second level of latent failure in the model is problems with supervision. Several of the themes identified in Table 11 directly relate to this factor.

Training. The review found 49 of the 1,021 cases listed some type of training issue as causal. If corporate and governmental oversight can be thought of as strategic limitations or guidance to aircrews, then implementing that guidance can be thought of as operational implementation of that strategy. One of the key ways guidance gets passed on to crews is through training. This training can be performed as initial training on new equipment or recurring training which is already required for maintaining pilot certification (Federal Aviation Agency, n.d.). Initial training deficiencies can be noted due to the operator's syllabus failing to address an issue or due to the manufacturer failing to provide the guidance to the operator:

> The complexities of the autothrottle and autopilot flight director systems that were inadequately described in Boeing's documentation and Asiana's pilot training, which increased the likelihood of mode error; ... the pilot flying's inadequate training on the planning and executing of visual approaches. (National Transportation Safety Board, n.d.-a, case 20130707X83745, Probable Cause Section, para. 1)

More common in the mishaps were examples where the information existed and was defined in the training syllabus, but for which the crew was not adequately trained in either complexity or recent experience. Such incidents could relate to anomalies already noted by other crews or design engineers (National Transportation Safety Board, n.d.-a, case 20110311X65739), system limitations exhibited when systems were performing with a downgraded capability (National Transportation Safety Board, n.d.-a, case 20070406X00376), or when individual crew members exhibited deficiencies that were not adequately addressed prior to being certified for flight (National Transportation Safety Board, n.d.-a, cases 20130814X15751 and 20130220X11432).

Experience. Another 33 cases listed some type of experience as causal. One of the mitigating factors available to operating airlines when new equipment is released or difficulties have been identified is to schedule crews with appropriate levels of experience to recognize and compensate for technological deficiencies. If the airline or its dispatchers fail to consider experience and then some technical malfunction occurs, the crew is at a disadvantage to handle the situation successfully. This situation was identified numerous times by the NTSB investigators (National Transportation Safety Board, n.d.-a, cases 20120607X54234, 20080925X0153, 20080925X01531, and 20140112X00750). Most of the cases cited used similar wording: "contributing to the accident were the pilot's lack of recent flight experience" (National Transportation Safety Board, n.d.-a, case 20101230X40808, Probable Cause Section, para. 1).

Supervision. There were 175 cases listed which showed some type of supervision as causal. As with any other job, flight crews require supervision to ensure adherence to guidance and to identify shortcomings before they become disruptive. Failure in this area was seen in multiple case that include both who is performing the supervision: "placed a pilot who inadequately emphasized safety in the position of company chief pilot" (National Transportation Safety Board, n.d.-a, case 20070613X00718, Final Narrative Section, para. 1), and the quality of the supervision being given: "inadequate surveillance by company personnel" (National Transportation Safety Board, n.d.-a, case 20040809X01167, Probable Cause Section, para. 1).

Preconditions for Unsafe Acts

The third level of latent failure under the theoretical framework is preconditions that lead to unsafe acts or do not prepare crews to deal with situations properly in flight. Three of the themes from Table 11 applied to this category.

Human factors. From the 1,021 cases, 68 listed some type of human factor as causal. One of the most commonly identified causes in all accidents (regardless of automation impact) is pilot error. Significant contributors to that type of error are human factor issues. It should therefore not be surprising to find that human factor considerations often intersect with automation concerns. Several of the most dangerous concerns identified by the investigators revolved around the "loss of situational awareness during the approach" (National Transportation Safety Board, n.d.-a, cases 20090715X83104, Probable Cause Section, para. 1), channelized attention on one system to the detriment of other systems such as "captain's preoccupation with the flight computer system" (National Transportation Safety Board, n.d.-a, case 20040513X0667, Probable Cause Section, para. 1), or "spatial disorientation while operating in dark night conditions" (National Transportation Safety Board, n.d.-a, case 20131120X00220, Probable Cause Section, para. 1).

Crew resource management. The review also found 37 cases listed some type of crew resource management issue as causal. In 2004, after lengthy study, the FAA imposed minimum training requirements for crew resource management (Federal Aviation Agency, 2004). However, the implementation has been left to the individual airline. Given the emphasis the FAA placed on CRM training, it is understandable to see it is often listed as a causal factor in mishaps by the NTSB (National Transportation Safety Board, n.d.-a, cases 20090425X65240, 20130814X15751, 20081003X16308, 20090127X92950, 20120908X34514, and 20120908X34514). Poor CRM manifested itself as "pilot's poor crew coordination and lack of cockpit discipline" (National Transportation Safety Board, n.d.-a, case 20080806X0116, Probable Cause Section, para. 1), or "the flight crew's nonstandard communication" (National Transportation Safety Board, n.d.-a, case 20130707X83745, Probable Cause Section, para. 1).

Fatigue. The largest group of precondition cases revealed that 228 cases listed some type of fatigue concern as causal. Both FAA and company guidance list minimal considerations to mitigate fatigue (Federal Aviation Agency, n.d.). This guidance may take the form of minimum periods devoted to crew rest prior to being authorized to fly on a mission, or in the form of maximum amount of time a crewmember may perform duties continuously to prevent overwork. Those limits are not tailored to individual crew members or non-standard situations. In those cases, it is ultimately up to the individual to determine if they are too fatigued to perform satisfactorily. Between personal motivations to continue on and perceived repercussions for taking oneself off the schedule and requiring a replacement, the fatigue of aircrew has often been cited as a mishap cause (National Transportation Safety Board, n.d.-a, cases 20130707X83745, 20130220X11432, and 20080806X0116). Fatigue could affect one member or the entire crew: "captain's performance deficiencies likely due to factors including, but not limited to fatigue…the first officer's fatigue due to acute sleep loss" (National Transportation Safety Board, n.d.-a, case 20130814X15751, Final Narrative Section, Para. 2).

Unsafe Acts

The final layer of failure in the Swiss Cheese model covers the actions of the crew during the flight. This is the last opportunity for a failure at any of the other levels to be recognized and corrected before it leads to a mishap. By the time an aircrew steps on the flight deck to conduct a passenger carrying flight, all of Reason's (1990) levels of latent failure have taken place. If those levels performed adequately, a failure that could end in catastrophe should have been caught. In the real world, either a failure could have occurred in one of those levels and not been caught, or the failure that causes a mishap could occur during the tactical execution of the flight.

Automated controls. The 1,021 cases in the qualitative analysis showed that 22 listed a specific type of automated control as causal. The introduction of new technology to the cockpit has indeed been identified by NTSB investigators as a causal factor in accidents for several different reasons. The first is the technology itself could be problematic. As a sample, the NTSB has identified "efforts to engage the autopilot" (National Transportation Safety Board, n.d.-a, case 20111130X22453, Probable Cause Section, para. 1), the "thrust reverser system, which permitted the failure of critical systems in the wheel well area to result in uncommanded forward thrust" (National Transportation Safety Board, n.d.-a, case 20081003X16308, Probable Cause Section, para. 1), "excessive maneuver in response to the traffic alert and collision avoidance system" (National Transportation Safety Board, n.d.-a, case 20091030X64143, Probable Cause Section, para. 1), "ineffective use of available information from their aircraft's electronic traffic advisory system" (National Transportation Safety Board, n.d.-a, case 20090808X42846, Probable Cause Section, para. 1), and "improperly used the navigation equipment, which had a graphical moving map display capability" (National Transportation Safety Board, n.d.-a, case 20051026X01740, Probable Cause Section, para. 1) as primary causes of mishaps. This non-exhaustive list does not include incidents where a piece of technology was considered secondary to another cause such as distraction.

Distraction. There were 58 cases in the final database which listed distraction as causal. Quite often investigators noted that either the complexity of a technological system itself or the effects of a malfunction in a system were causal to mishaps. One common effect of technology that was seen to cause problems was distraction. In some cases a system performing within its parameters was a cause in itself, such as normally "programming the global positioning system receiver" (National Transportation Safety Board, n.d.-a, case 20111102X13005, Probable Cause Section, para. 1). In other scenarios, a malfunction in the system caused distraction and a subsequent mishap, even if the malfunction itself was not a threat to safe flight such as the "pilot's distraction by the reported malfunction of the autopilot system" (National Transportation Safety Board, n.d.-a, case 20131202X13615, Probable Cause Section, para. 1).

Overreliance. In 18 of the cases analyzed, pilot overreliance on automation was listed as causal. Technology and automation have been incorporated into modern cockpits long enough that many pilots either do not have experience without the technology or have come to depend on it to the exclusion of the traditional pilot role. The NTSB has found many instances of the situation such as the "pilot's overreliance on the autopilot system and his inability to hand-fly the airplane once the autopilot was disconnected" (National Transportation Safety Board, n.d.-a, case 20121007X94725, Probable Cause Section, para. 1), or the "pilot's reliance on the on-board engine-monitor instrument instead of the fuel quantity gauges" (National Transportation Safety Board, n.d.-a, case 20090404X04940, Probable Cause Section, para. 1).

Slow/improper response. For the final coded theme of the qualitative analysis, 103 cases listed some slow or improper pilot actions as causal. Another impact of automation that investigators found in these cases was a tendency for pilots to delay taking appropriate action based on the real or perceived assessment that the technology would intervene. In these cases the slow or improper response led the pilot to delay taking action that would have prevented the mishap until the window of opportunity to do so had past. In such cases, the pilot may have not verified "the airplane's flight management computer was properly configured for takeoff and the captain failed to perform the correct action in response" (National Transportation Safety Board, n.d.-a, case 20140314X21725, Probable Cause Section, para. 2), or did not have sufficient understanding of how a system worked, as when "lack of familiarity with the airplane's autobrake system distracted them from thrust reverser usage during the challenging landing" (National Transportation Safety Board, n.d.-a, case 20051213X01964, Probable Cause Section, para. 1).

Research Questions

Following the third cycle determination of themes, the results were reviewed for their applicability to each of the research questions. The mapping of themes to questions proved to form relatively clear relationships.

Does the Implementation of Automation Technology Have a Significant Impact on Incidence of Aircraft Mishaps in United States Aircraft with Respect to Human Factor Issues, Technological Deficiencies, and Human-Machine Interfaces

Both the quantitative and qualitative data indicate that technology does have a small but significant effect on the incidence of aircraft mishaps. In the ordinal regression analysis, the contribution of technology to accidents was shown to have low predictive capability in the severity of mishap, but multiple factors were shown to be significant. Given the high safety rate in the aviation industry, it is likely that improvements in safety will only be found in small increments as there is no major factor that is easily addressed as there was in the early days of aviation when the reliability of the equipment was suspect (Koeppen, 2012). In the qualitative analysis, multiple factors related to technology and human interaction with technology presented themselves as themes from actual mishap investigations. It is from the investigator and pilot narratives of the causal factors that the most valuable information may be drawn with respect to future actions to mitigate risk.

Impact of technological deficiencies. The ordinal regression showed that almost all categories of technology examined presented a significant factor in the outcome of mishaps and were seen at the organizational influence and unsafe acts levels of the Swiss Cheese model. The aviation database, the use of automated controls, the interface of technology with the pilot, the use of digital communications systems, and the flight computer all showed an impact on the outcomes. Although flight management systems did not show significance, that is not a mitigating factor for the other types of technology. On the contrary, it is possible that if any effects of the FMS were actually significant, they may have been attributed to the systems they interact with: the pilot interface or the flight computer. Whether the FMS does or does not have a significant contribution, the evidence of technological issues on the flight deck is substantial.

Technological deficiencies of both the equipment and the ability to use it were shown in many NTSB investigations. These deficiencies covered a wide spectrum. Pilots may have attempted to use the equipment, but incorrectly as in the case where the "airplane was equipped with a global positioning system (GPS) that was tied into a moving map cockpit display, which included terrain features and obstacles," (National Transportation Safety Board, n.d.-a, case 20051026x01740, Final Narrative Section, para. 1) but resulted in impact with a mountain in spite of this equipment. There may also have been technological limitations on the equipment from the sheer volume of data available, such as "the flight crew did not recognize or respond to the enhanced ground proximity warning system (EGPWS) warning, which alerted because the EGPWS did not recognize the runway since it was less than 3,500 feet long" (National Transportation Safety Board, n.d.-a, case 20060725x01011, Final Narrative Section, para. 4).

Further technological issues have been identified. These include limitations on the equipment's ability to respond predictably when receiving inadequate input information, such as "the flight crew's failure to properly configure and verify the flight management computer for the profile approach" (National Transportation Safety Board, n.d.-a, case 20130814x15751, Probable Cause Section, para. 1). In such cases the initial cause may have been something else, such as a pilot mistake, but the mistake could not be caught before disaster because the technology acted in a manner consistent with it functioning normally. If some failure mode or warning had been given, there is a possibility the crew would have caught their mistake prior to the accident. On the opposite end of the spectrum, incidents were seen where the technology warning signal either came so late in the process or startled the crew at an inappropriate time that it caused or contributed to the accident itself. One example was when "1.5 seconds after the traffic alert and collision avoidance system (TCAS) resolution advisory (RA) occurred, the captain initiated a series of excessive control inputs which resulted in a positive vertical acceleration of approximately 1.6g" (National Transportation Safety Board, n.d.-a, case 20091030x64143, Final Narrative Section, para. 1) which resulted in one serious and one minor injury.

Impact of human factor issues. In a review of all the narrative mishaps, one of the most commonly cited causes was some form of human factor issue and these generally presented in either the unsafe supervision, the preconditions for unsafe acts, or the unsafe acts level of the Swiss Cheese model. Human factors issues have been studied extensively (Amalberti, 2009; Chandra & Grayhem, 2012; Endsley, 1996) as a field in itself, and in relation to aviation. However, the analysis shows that there is still room for specialists to look at the remaining human factor causes of mishaps with an eye toward reducing or mitigating them.

From the analysis, human factor issues presented themselves in a variety of ways including distraction, overreliance on automation, channelized attention, fatigue, confusion, or poor crew resource management. Some of these issues resulted from inadequate supervisory decisions such as "mismanagement of an abnormal flight control situation...which placed a pilot who inadequately emphasized safety in the position of company chief pilot and check airman" (National Transportation Safety Board, n.d.-a, case 20070613x00750, Probable Cause Section, para. 1).

Human factor issues cover a broad range of issues, each with potentially different causes or avenues of mitigation. Some relate to loss of situational awareness (National Transportation Safety Board, n.d.-a,c ase 20090715x83104) or some subcategory of that such as spatial disorientation: "airplane experienced failures of flight/navigation instruments. The 1,050-hour pilot was attempting to maneuver the airplane using partial panel techniques. Radar data showed that the pilot appeared to be experiencing spatial disorientation" (National Transportation Safety Board, n.d.-a, case 20040106x00018, Final Narrative Section, para. 1).

Human limitations also showed themselves as distraction: "the pilots' inattention to basic aircraft control while attempting to program the autopilot system" (National Transportation Safety Board, n.d.-a, case 20090428x81708, Probable Cause Section, para. 1). Other manifestations included "the pilot's inability to hand-fly the airplane once the autopilot was disconnected" (National Transportation Safety Board, n.d.-a, case 20121007x94725, Probable Cause Section, para. 1) or "the captain's preoccupation with the flight management system" (National Transportation Safety Board, n.d.-a, case 20120513x00667, Probable Cause Section, para. 1).

Another commonly mentioned human factors issue was crew resource management. CRM showed itself in unusual intra-cockpit communication: "nonstandard communication and coordination regarding the use of the autothrottle and autopilot flight director systems" (National Transportation Safety Board, n.d.-a, case 20130707x83745,

Probable Cause Section, para. 1) or unexpected actions by a crewmember: "the captain's failure to communicate his intentions to the first officer once it became apparent the vertical profile was not captured" (National Transportation Safety Board, n.d.-a, case 20130814x15751, Probable Cause Section, para. 1).

Impact of human-machine interfaces. Closely related to the human factor issues above is how the human in the system interacts with the technology. In many cases, the diverted attention while interfacing with the automation is problematic, as shown in the example above. In many more cases, the confusion a pilot encounters while interfacing with technology during an otherwise minor malfunction leads to a major problem. Examples of this phenomena are seen in " it is likely that the pilot became focused on the autopilot system and diagnosing the reported problem" (National Transportation Safety Board, n.d.-a, case 20131202x13615, Final Narrative Section, para. 1) or "the pilot's distraction with the inappropriately set transponder" (National Transportation Safety Board, n.d.-a, case 20050805x01171, Probable Cause Section, para. 1).

Furthermore, these types of human-machine interfaces could occur in the preconditions for unsafe acts level as well as the unsafe acts level of the Swiss Cheese model. Examples of the former are seen in the "distraction with programming a global positioning system receiver" (National Transportation Safety Board, n.d.-a, case 20111102x13005, Probable Cause Section, para. 1) or distraction "while attempting to program the autopilot system" (National Transportation Safety Board, n.d.-a, case 20090428x81708, Probable Cause Section, para. 1).

Is the Current Policy Guidance on Cockpit Automation Sufficient with Respect to Scope, Relevance, Compliance, and Enforcement

The impact of policy guidance can be seen at all levels of the Swiss Cheese model. In the organizational influences, both company policies and government oversight play a role in how aircraft are certified, staffed, and scheduled as well as how pilots are approved, supervised, and evaluated. At the unsafe supervision level, how supervisors control and train pilots are critical factors. At the preconditions for unsafe acts, the pilot's level of experience and internalized knowledge play a role. And at the unsafe acts level, whether pilots adhere to the guidance and how they implement it has been seen to vary.

Scope of guidance for passenger carrying aircraft. The guidance and certification requirements for technology in passenger carrying aircraft occur primarily at the organizational influence level. The policies that a company publishes can determine what type of technology will be purchased and the rules for its use by crews. The certification of the technology is a function of the government policy and the government is also responsible for follow-up inspections to ensure equipment is installed properly, crews are trained appropriately, and the equipment used by crew is within the scope of authorization.

In some cases, investigators found these responsibilities overlapped and failures could be attributed to both the government and the airline operator, as in:

> Operator's failure to require its flight crews to file and to fly a defined route of flight, the operator's failure to ensure that the flight crews adhered to company policies and FAA and DoD federal safety regulations, and the lack of in-country oversight by the FAA. (National Transportation Safety Board, n.d.-a, case 20041213x01969, Probable Cause Section, para. 1)

Or in some cases the operator failed their duty and the government failed in enforcement:

> Contributing to the accident was the operator's inadequate procedures for operational control and flight release and its inadequate training and oversight of operational control personnel. Also contributing to the accident was the Federal Aviation Administration's failure to hold the operator accountable for correcting known operational deficiencies and ensure compliance with its operational control procedures. (National Transportation Safety Board, n.d.-a, case 20131126x92315, Probable Cause Section, para. 1)

However, it was also the case that one or the other of the groups tasked with overall guidance would fall short. This could be the government providing inadequate guidance: "the Federal Aviation Administration's failure to establish adequate certification requirements for flight into icing conditions, which led to the inadequate stall warning margin provided by the airplane's stall warning system" (National Transportation Safety Board, n.d.-a, case 20050304x00167, Probable Cause Section, para. 1). Or the fault could be placed with the airline's own policies:

> Southwest Airline's 1) failure to provide its pilots with clear and consistent guidance and training regarding company policies and procedures related to arrival landing distance calculations; 2) programming and design of its onboard performance computer, which did not present inherent assumptions in the program critical to pilot decision making; 3) plan to implement new autobrake procedures without a familiarization period; and 4) failure to include a margin of safety in the arrival assessment to account for operational uncertainties. (National Transportation Safety Board, n.d.-a, case 20051213x01964, Probable Cause Section, para. 1)

Relevance of guidance. What guidance is published was found to be lacking in some cases relative to the change in technology. This deficiency could be seen at the Swiss Cheese organizational failure level when government or company policy was not sufficient for the current technology, or it could be seen at the unsafe supervision level in inadequate training programs, or it could be seen in the preconditions for unsafe acts

level with unacceptable pilot preparation.

At the organizational level, deficiencies could be seen in companies failure to

provide a "program or policy within the company operations specifications or other

manual for higher approach minimum limitations based upon experience for company

pilots of piston engine powered airplanes such as the accident airplane" (National

Transportation Safety Board, n.d.-a, case 20140112x00750, Final Narrative Section, para.

3). Whereas government guidance could be seen to fall short in certification requirements

when investigators faulted:

> Deficiencies in Learjet's design of and the Federal Aviation Administration's
> (FAA) certification of the Learjet Model 60's thrust reverser system, which
> permitted the failure of critical systems in the wheel well area to result in
> uncommanded forward thrust that increased the severity of the accident; (2) the
> inadequacy of Learjet's safety analysis and the FAA's review of it, which failed
> to detect and correct the thrust reverser and wheel well design deficiencies after a
> 2001 uncommanded forward thrust accident. (National Transportation Safety
> Board, n.d.-a, case 20081003x16308, Probable Cause Section, para. 1)

Training issues with relevance were also noted by the NTSB when they found the

pilot "seemed very disoriented with the new technology on this flight and previous

flights" (National Transportation Safety Board, n.d.-a, case 20110311x65739, Final

Narrative Section, para. 2) or when they noted "operator's inadequate training of the

pilot...in the make and model of the airplane" (National Transportation Safety Board,

n.d.-a, case 20140112x00750, Probable Cause Section, para. 1). Lack of relevant

guidance could also be seen as individual pilots failing to avail themselves of the

necessarily knowledge prior to flight or the "pilot's lack of familiarity with the airplane's navigational equipment" (National Transportation Safety Board, n.d.-a, case 20090715x83104, Probable Cause Section, para. 1).

Compliance and enforcement of guidance. Regardless of the efficacy of the available guidance, it is insufficient if it is not followed and compliance and enforcement of guidance generally falls within the responsibility of the organizational failure or unsafe supervision levels of Reason's (1990) model. Investigators found mishaps they could directly attribute to a lack of either government enforcement of standards or company enforcement of compliance.

The government's shortcomings could be seen in several narratives including the "FAA Principal Operation Inspector's (POI) failure to complete the appropriate form" (National Transportation Safety Board, n.d.-a, case 20080925x01531, Final Narrative Section, para. 2) or "the failure of the Federal Aviation Administration to require crew resource management training and standard operating procedures for Part 135 operators" (National Transportation Safety Board, n.d.-a, case 20080806x01166, Probable Cause Section, para. 1).

Airline failings were seen in mishaps such as the "certificate holder's loss of operational control" (National Transportation Safety Board, n.d.-a, case 20080925x01531, Probable Cause Section, para. 1) or the "deficiencies in the NMSP aviation section's safety-related policies" (National Transportation Safety Board, n.d.-a, case 20090610x23159, Probable Cause Section, para. 1). In some cases, although adequate guidance and compliance policies were in place, they failed in execution at the supervision level as when "inadequate surveillance by company personnel" (National

Transportation Safety Board, n.d.-a, case 20040809x011667, Probable Cause Section, para. 1).

Does the Type of Training Received by Cockpit Crew and the Level of Experience with Cockpit Automation Change the Relationship of Flight Deck Automation and Aircraft Mishaps

Crew training and experience are factors that cut across all levels of the Swiss Cheese model. At the organizational influences level, the type and quantity of training is determined. At the supervision level, the training is actually provided and experience requirements for the flight crew are established. At the preconditions level, the training is received and the pilot internalizes the information. And at the acts level, the training and experience can either contribute to resolving an issue or exacerbate the problem.

Some of the training concerns raised by the investigators included airline policies toward training such as the operator's "inadequate training and oversight of operational control personnel" (National Transportation Safety Board, n.d.-a, case 20111130x23954, Probable Cause Section, para. 1) or failure to provide training "commensurate to the pilot's experience" (National Transportation Safety Board, n.d.-a, case 20140112x00750, Probable Cause Section, para. 1). Other corporate training issues appeared at the supervisory level when training was provided to pilots such as when the NTSB found within the same mishap "complexities of the autothrottle and autopilot flight director systems that were inadequately described in Boeing's documentation and Asiana's pilot training, which increased the likelihood of mode error" and "the pilot flying's inadequate training on the planning and executing of visual approaches" (National Transportation Safety Board, n.d.-a, case 20130707x83745, Probable Cause Section, para. 1).

At the preconditions and unsafe acts levels, multiple findings appeared in the investigations. Inadequate preparation of a pilot could lead to problems such as when faced with "pilot's decision to perform a test flight on a system for which he lacked a complete working knowledge" (National Transportation Safety Board, n.d.-a, case 20110311x65739, Probable Cause Section, para. 1), or when the pilot's "training program was fragmented over approximately ten months, and while in accordance with FAA regulations, may have adversely affected his consolidation of skills and experience" (National Transportation Safety Board, n.d.-a, case 20090507x00926, Final Narrative Section, para. 4). Furthermore, even adequately trained crews were seen to make errors that were attributed to either a lack of total experience or a lack of experience specific to the automation they were using on that flight. Such examples are seen in an airline's "plan to implement new autobrake procedures without a familiarization period" (National Transportation Safety Board, n.d.-a, case 20051213x01964, Probable Cause Section, para. 1) or the pilot's "lack of experience in the type" (National Transportation Safety Board, n.d.-a, case 20080925x01531, Probable Cause Section, para. 1).

What are the Policy Actions that have Shown and Ability to Mitigate Negative Effects or Enhance Positive Effects of Flight Deck Automation on Aviation Safety

Both human factors research and international experience have shown aviation policies that have shown promise in mitigating negative effects and enhancing positive effects of automation, and those are discussed later in this study. For the analysis of

NTSB reports, the investigators commonly identified either a lack of a policy or a lack of adherence to a policy that might have mitigated the impact had it been published and/or complied with. Some of the more common observations are listed here.

The first item investigators often mentioned was crew resource management. Although CRM guidance from the FAA has been around since at least 2004, it is generally left to the airline to develop the actual training program. This has led to some shortfalls that the investigators identified. For example, while the Captain of a flight was overly focused on the weather radar in poor weather conditions,

> The First Officer, who was flying at the time, had asked the Captain about ten minutes prior to the impact if their altitude was high enough to clear the upcoming terrain, but the Captain did not respond, and the First Officer did not challenge the Captain about the issue. (National Transportation Safety Board, n.d.-a, case 20090425x65240, Final Narrative Section, para. 1).

Other situations where CRM was identified as a policy that could have been constructive if properly trained and followed was when "The copilot completed the approach and landing checklist items but did not call out items because the PIC preferred that copilots complete checklists quietly" (National Transportation Safety Board, n.d.-a, case 20101001x13003, Final Narrative Section, para. 2) and "the pilots exhibited poor crew resource management by not using the appropriate chart for the contaminated runway, not recognizing the runway was too short based on the conditions, failing to reset their airspeed bugs before the approach" (National Transportation Safety Board, n.d.-a, case 20120918x34514, Final Narrative Section, para. 4).

More generally, human factors issues were identified as policies that could be improved by the investigators. They often identified situational awareness or spatial disorientation as a significant factor, and in some cases, the only factor "examination of the wreckage found no evidence of any mechanical malfunction" (National Transportation Safety Board, n.d.-a, case 20040300x00399, Final Narrative Section, para. 1) or "no anomalies were noted with the gyro instruments, engine assembly or accessories" (National Transportation Safety Board, n.d.-a, case 20080129x00119, Final Narrative Section, para. 1).

Another human factor problem that exposed a policy deficiency was distraction due to the technology. This distraction could be due to difficulty with normal operations such as when "The pilot then realized that while he was trying to engage the autopilot, the airplane's heading had drifted and the airplane was headed toward rapidly rising terrain" (National Transportation Safety Board, n.d.-a, case 20111130x22453, Final Narrative Section, para. 1) or "the pilot failed to properly engage the autopilot altitude preselect mode; the altitude hold mode was entered instead" (National Transportation Safety Board, n.d.-a, case 20090428x81708, Final Narrative Section, para. 2). Alternatively, the distraction could be due to a malfunction such as when "the pilot became focused on the autopilot system and diagnosing the reported problem" (National Transportation Safety Board, n.d.-a, case 20131202x13615, Final Narrative Section, para. 1) or "The electrical anomalies likely distracted the pilot and led to his subsequent loss of airplane control" (National Transportation Safety Board, n.d.-a, case 20120421x43732, Final Narrative Section, para. 3).

Another area of human factor influence that policy actions might impact is the area of fatigue. The NTSB explicitly identified impacts of fatigue as causal despite the pilots conforming to existing sleep and work-cycle requirements:

> Fatigue associated with sleep loss, circadian disruption, and long duty hours can lead to increased difficulty in sustaining and directing attention, memory errors, and resultant lapses in performance. An NTSB safety study found that flight crewmembers who were awake for more than 12 hours made more procedural errors, tactical decision errors, and errors of omission than those awake less than 12 hours. (National Transportation Safety Board, n.d.-a, case 20130220x11432, Final Narrative Section, para. 8)

Beyond the sleep/work cycle for any given day, the cumulative effect of fatigue was identified when the pilot "experienced a demanding 10-day trip schedule prior to the incident involving multiple time zone crossings and several long duty periods, and reported difficulties sleeping prior to the accident leg" (National Transportation Safety Board, n.d.-a, case 20090507x00926, Final Narrative Section, para. 3).

Areas investigators identified as beneficial policies were not limited to the aircrews. Stronger enforcement policies by the government were noted:

> Although FAA inspectors were providing surveillance and noting discrepancies within the company's procedures and processes, the FAA did not hold the operator sufficiently accountable for correcting the types of operational deficiencies evident in this accident, such as the operator's failure to comply with its operations specifications, operations training manual, and GOM [General Operations Manual] and applicable federal regulations. (National Transportation Safety Board, n.d.-a, case 20131126x92315, Final Narrative Section, para. 1)

Also noted as beneficial by negative case analysis was the enforcement of standards by the FAA:

At least one inspector from the FAA's Flight Standard District Office (FSDO) was aware of the history of improper exercise of operational control of flights by the airplane owner (not the certificate holder) and no action was taken to stop this practice. Additionally, the FAA FSDO surveillance records revealed that the certificate holder was rated as satisfactory with no comments noted, although throughout the investigation numerous discrepancies were found that within the company that did not comply. (National Transportation Safety Board, n.d.-a, case 20080925x01531, Final Narrative Section, para. 2)

Finally, the investigators often took care to specify training programs could be both beneficial when performed properly or detrimental when incomplete or poorly prepared.

On the beneficial side, they noted that the pilot

Had received formal training in two models of airplanes equipped with the same flight display system, and also experienced a failure of the display system on a previous airplane. Thus, the pilot was likely familiarized with the functionality of the PFD [Primary Flight Display]. (National Transportation Safety Board, n.d.-a, case 20070406x00376, Final Narrative Section, para. 5)

But they also noted that poor training could lead to technology mistakes such as when "although the pilot had received antiskid system failure training during his recurrent simulator training on January 4, 2013, he stated in post-accident interviews that he did not think they needed the antiskid system for the landing" (National Transportation Safety Board, n.d.-a, case 20130220x11432, Final Narrative Section, para. 6).

CHAPTER V

DISCUSSION AND CONCLUSION

The purpose of this study was to contribute to the development of United States aviation policy with regard to automation technology on passenger carrying aircraft in the National Airspace System. The primary finding of this analysis was that automation technology does appear to have an impact on aviation safety. This result provides an avenue for regulators to improve safety by investigating policies that might mitigate the impact or enhance positive factors.

When the first aircraft accident was primarily attributed to improper usage of automation technology in Cali, Colombia in 1996 (Simmon, 1998), there was a heightened emphasis on ways to mitigate the potentially negative safety impacts of human interaction with automation while maintaining the recognized benefits in efficiency and safety that technology can bring. The FAA considered the issue significant enough to require airline pilots to receive crew resource management training in 2004 which included dealing with automation (Federal Aviation Agency, 2004). However, the FAA's position has traditionally been one of setting general guidance rather than detailed requirements. Also since that time, the world has become more comfortable with technology in many aspects of daily life and the pilots flying today's aircraft are immersed in that culture. Therefore, there is a concern that the hard-learned lessons from past interactions with technology are being marginalized at the same time that technological innovations are rapidly being incorporated into aircraft flight decks.

The literature to date has shown that automation has been a part of aviation from almost the beginning and has taken on a much more integral role as the volume of flights, the number of passengers, and the technological capabilities have developed. But while technology has helped improve efficiency of air traffic, the complexity of that on technology has caused concerns, especially with respect to human factor issues related to human-machine interactions, the changing role of the pilot within the cockpit ecosystem, and automation induced problems. As the FAA plans a major overhaul of the United States airspace, the concerns are taking on an increased sense of urgency. Under the theoretical framework of the Swiss Cheese model developed by Reason (1990) and expanded by Shappell and Wiegmann (2000) the publicly reported aircraft mishaps in the ASRS and NTSB were examined to determine the impact of technology.

The study was conducted using an explanatory sequential mixed method design using elemental methods (Saldaña, 2013) derived from the epistemological nature of the research questions. A pragmatic approach to the narratives was used where the descriptive factors were analyzed for patterns to capture the essence of the causal factors and map them to concepts that formed emergent themes. These themes naturally grouped into the four levels of the Swiss Cheese model and were used to answer the research questions that drove the study.

Research Questions

Following the quantitative and qualitative analysis, the resulting patterns and themes were applied to the study's research questions. From this, several conclusions were drawn.

Does the Implementation of Automation Technology Have a Significant Impact on Incidence of Aircraft Mishaps in United States Aircraft with Respect to Human Factor Issues, Technological Deficiencies, and Human-Machine Interfaces

The quantitative analysis was performed to identify factors of significance which were then used to focus the qualitative analysis under the sequential concept of the methodology. The results showed that technology did have an impact.

Quantitative. An ordinal regression was used to model the severity of outcomes ranked as minor injury, serious injury, or fatality while looking at various types of flight deck automation as independent, predictor variables. The predictors are registered as dichotomous effects, they are either cited in the causal narrative or they are not. While the model was a weak predictor of outcomes, it did show significance in specific types of automation, specifically, the aircraft navigation database, the automated controls such as autopilot and autothrottles, the interface for pilot input and feedback, the flight computer, and digital communication systems. These results confirmed the need for further analysis and provided the focus for the follow on qualitative analysis.

Qualitative. Using the results of the quantitative analysis and the theoretical taxonomy of Jones et al (2013) as a starting point, an iterative coding process was applied to the subset of 1,021 NTSB narratives that fit the criteria of post 2004 FAA directed CRM training, some level of injury, and one of the factors identified in the quantitative analysis. In the first and second cycles of coding, structural, descriptive, and in-vivo elemental methods (Saldaña, 2013) were identified from the causal finding section of the NTSB mishap investigations. After reviewing all the codes a third coding cycle was performed in which the descriptive codes were grouped into patterns to identify themes for conclusions. Twelve themes related to how automation considerations impact flight deck operations and aircraft safety presented during this process: company policies, governmental oversight, training issues, insufficient experience levels, poor supervision, human factors issues, crew resource management deficiencies, fatigue, control of the aircraft being delegated to automation, crew distraction, overreliance on automation, and slow or improper responses. These themes led to the conclusion that automation technology, how humans interact with the automation, or human factor issues related to the use of automation create a potentially hazardous situation that may be suitable for policy actions designed to mitigate the risk.

Is the Current Policy Guidance on Cockpit Automation Sufficient with Respect to Scope, Relevance, Compliance, and Enforcement

Because the safety of air travel relative to other forms of transportation is well known (International Civil Aviation Organization, 2013), the current level of policy guidance can be considered adequate when viewed on itself. However, if the purpose of the government and airlines is to increase safety even more, then the causes identified in question 1 provide an opportunity for policy improvement within those areas.

Scope. While every pilot is likely to want to improve the safety of the aircraft they are flying if for no other reason than self-preservation, this study is interested in raising the level of safety for the traveling public. The majority of aircraft passengers are flying on commercial aircraft and the oversight for those flights is the responsibility of the FAA and the airline operating the aircraft.

Relevance. During the analysis, both FAA regulatory guidance and airline operating procedures were seen to be insufficient for the rapidly developing pace of technology. Furthermore, the certification process the government uses to approve new equipment on aircraft presents difficulties in both keeping pace with technological advances and adds significant cost that reduces the incentives for airlines to adopt new equipment for only the purpose of improving crew performance. Training was also seen to be an area of potential improvement as the training syllabi used by airlines was not standardized, tended to lag the introduction of new technology in addressing negative effects, and was often provided at the minimally required level, leaving pilots with weak spots in their capabilities.

Enforcement. Even in cases where existing policy was deemed to be sufficient, a related area for improvement was the enforcement of the policy guidance. This was another area where responsibility was shared by the FAA and the airline. In some cases, the government inspection program failed to identify poor airline procedures or situations where pilots failed to comply with regulations. Within an airline, management had a responsibility to ensure compliance with government requirements and with company standard operating procedures, but was sometimes seen to fall short in that role.

Does the Type of Training Received by Cockpit Crew and the Level of Experience with Cockpit Automation Change the Relationship of Flight Deck Automation and Aircraft Mishaps

Within the NTSB investigations, training and experience were cited nearly 10% of the time for direct concerns such training on the use of technology, length of training time, crew resource management training, and experience with the type of equipment being used. Beyond the direct issues referenced, many of the causal narratives identified factors that lend themselves to training for improvement. For example, if adequate crew rest time was provided, but the pilot used that time inefficiently, training on circadian rhythm or time management might be effective. In addition, experience was noted as a significant issue in general, but looking into the issue further showed that total experience was often not sufficient in itself. Rather, experience with the specific equipment being used was required. Due to budgetary considerations and advancing technology over time, many airlines have fleets with widely varying technological capabilities. On any given day a pilot with the proper certifications might fly in a brand new 737-900 type aircraft with the most modern equipment available, and the next day the same pilot might be scheduled to fly a 10-year old 737-800 model. If one thinks of the changes in technology in just their own experience over the last 10 years—cell phones, computers, television, etc.—it is easy to understand the increased effort needed by the pilot. That crew member needs to be aware of the capabilities and limitations of the equipment they are faced with, and they need to be able to do so when faced with the most stressful and time-critical situations.

What are the Policy Actions that have Shown and Ability to Mitigate Negative Effects or Enhance Positive Effects of Flight Deck Automation on Aviation Safety

Because one of the charters of the NTSB when they perform an investigation is to look for recommendations they can make for the future that might have prevented the mishap they are investigating (National Transportation Safety Board, n.d.-c), the body of the expert narratives often provide insight into what policies worked and what ones did

not. Under this line of reasoning, several areas clearly appeared as significant from the

investigators. First, crew resource management training was seen as a positive factor,

though often specific areas were identified that could be added to the syllabus for

improvement. Another factor often seen was failure to include an adequate human factor

consideration into the design and certification of the equipment. Making equipment easier

to interact with, to interpret, to control, and to override when necessary could mitigate

limitations of the human being suddenly thrown into the system when unplanned events

occur.

Related to the human factors issue, but viewing the concern from a different

perspective was the addition of external means of mitigation. For example, spatial

disorientation and loss of situational awareness were seen in a relatively large number of

cases. In these situations, external hardware or the introduction of more intuitive cockpit

displays might reduce the incidence. Using a third party on the ground to monitor the

aircraft via ground-based radar or space-based satellites also has the potential to provide

an extra margin of safety until the pilot regains their faculties and can address the

situation.

Collective Findings and Future Research

Both the quantitative and qualitative analysis showed evidence of automation

technology having an impact on aviation safety. Although it is beyond the scope of this

paper to look at the efficacy of any specific policy action that might be implemented,

viewing the results under the prism of the Swiss Cheese theory (Reason, 1990;

Wiegmann & Shappell, 2001) provides a road map for future consideration. In an environment of limited time and financial resources, focusing attention on those areas most susceptible to improvement provides both a higher probability of success and a higher probability of obtaining the needed resources to implement.

Findings in the Theoretical Context

In the view of the Swiss Cheese Model, the myriad of actions necessary to prepare and execute a passenger carrying flight pass through four levels which may filter out errors and prevent a mishap. For a mishap to occur, the error has to pass through holes or deficiencies in all of the levels. At the top level, latent organizational influences were shown to have weak spots in the areas of company policy, governmental oversight, and governmental certification. At the next latent level, supervision was seen to have areas for improvement in the type and volume of training provided to air crew members, the scheduling and dispatch of specific crews on specific models of aircraft, and on monitoring of individual pilots. The next level of potential failure is also latent to the activity and is seen in the preconditions necessary for crews to perform unsafely. Examples of this were seen in inadequate crew preparation for crew resource management techniques, insufficient exposure to human factor issues and corrective actions, and poor fatigue management techniques. A flight does not occur until all the previous levels of potential failure have occurred. Once the crew initiates a given flight, they represent the last line of defense to catch any errors and provide an opportunity for an active failure by performing unsafe acts. The types of acts seen in the NTSB reports included turning over too much authority for aircraft control to automation equipment or doing so at an inappropriate time, distraction from primary control of the aircraft due to automation feedback or confusion, overreliance on the use of automation to the detriment of pilot skills and timely control intervention, and slow or improper responses to unexpected events due to complacency or lack of knowledge of the system.

Future Research

All the areas found in this study present a potential opportunity to improve aviation safety. Some will be more promising than others in both efficacy and economic viability. In order to determine which actions to pursue, experts in the appropriate fields should analyze potential policy recommendations and provide feedback on the usefulness of each. Fortunately entire fields of study have already been established with the capability to make this evaluation. Organizations such as the Department of Transportation's Volpe Center (Chandra & Kendra, 2010) are already engaged in human factor research relative to aviation. Aeronautical and system engineers already have the expertise to implement changes into hardware or bring new technologies to bear on existing problems. Domestic and international organizations such as the Air Transport Association, the ICAO, and the European Aviation Safety Agency are already interested in many of the same issues (European Aviation Safety Agency, n.d.; International Air Transport Association, n.d.; International Civil Aviation Organization, 2013). By either capitalizing on their existing research or partnering with them for future research, the NAS can be made safer as well as the corresponding international system. The policy recommendations below would provide a good starting point for evaluation, which can then be updated or modified by the expertise of researchers in the particular field of interest.

Future Policy Considerations

Although it was beyond the scope of this study to evaluate specific policy proposals, a close evaluation of the narrative information revealed several specific proposals worthy of consideration. For those recommendations that the FAA determine to have potential, qualified researchers and experts should evaluate the impacts and perform a cost-benefit analysis for inclusion into federal guidance. Additionally, where appropriate, airlines may choose to implement some of these recommendations on their own initiative after performing their own benefit analysis. If airlines choose to implement any recommendations, the resulting change in outcomes would provide beneficial feedback to the FAA in considering future regulations

Organizational Recommendations

- FAA certification procedures could be revised to include a stronger emphasis on human factors considerations. The FAA Order 9550.8, Human Factors Policy (Federal Aviation Agency, 1993) has not been updated since 1993. The state of the art has advanced considerably in that time as has the capability of technology.

- Airline purchasing decision related to newly developed technology could include an explicit requirement for human factors testing with emphasis on the functions the airline intends to authorize for use as well as the distractions associated with the functions they do not plan to authorize.

- The FAA could develop and maintain a specific crew resource management syllabus. Doing so would standardize training across airlines and would allow for updates due to technological changes to be incorporated more quickly than the current process. Under current procedures, individual airlines must update the training against a strong disincentive due to cost considerations. Alternatively this initiative could be delegated to an industry group such as the Air Transport Association, but backed by the regulatory authority of the FAA.

- Both FAA and airline internal inspection programs could be strengthened and expanded to ensure that airlines are in compliance with FARs, procedural guidance short of regulations, and minimal best practices. Inspection programs could also increase spot checks of aircrews for compliance.

- The FAA could evaluate the benefit of subsidizing equipment upgrades where they would have the greatest impact. For example, the natural business cycle of the airline industry results in diverse fleets of aircraft in both age and capability. For aircraft that may be 20 years or older, the limited functionality on the flight deck compared to modern aircraft is a drag on the entire NAS. By subsidizing upgrades of older aircraft to modern reporting and communication capabilities, ground based controllers will increase the flexibility to position aircraft in the safest configuration possible.

Supervisory Recommendations

- Training syllabi could be updated to increase the exposure to failure modes in available technology, increase the amount of time spent on pilot skills such as hands-on flight control, and recognition of uncommon events relating to technology. Training could be expanded in both recurrent simulator training as well as real-time training during flight when the crew composition includes a certified instructor pilot.

- Dispatch procedures could be updated to ensure minimal experience levels are achieved across the crew complement that are specific to the type of equipment on board. For example, merely ensuring a minimum crew experience in an Airbus 320 does not ensure that the crew has relevant experience for the version of the plane manufactured in the 1990s as the ones manufactured in the 2010s.

Precondition Recommendations

- Anti-fatigue rules could be updated to include not only maximum work day and minimum rest periods, but to also include considerations of multi-day schedules and time zone changes that do not allow circadian rhythms to adjust.

- New external technology could be brought to bear to include third-party monitoring of aircraft. Technology such as Automatic Dependent Surveillance – Broadcasting (ADS-B) provided real-time, precision situational awareness (Federal Aviation Agency, 2011) and is already mandated for certain aircraft by 2020. This mandate could be expanded to a larger segment of the aviation community or to include ADS-C (Contract) which allows for two-way digital exchange of information instead of the one-way exchange in ADS-B.

Operational Recommendations

- Warning methodologies and failure modes could be made more transparent to the flight crew. If a standardized message or display was provided across aircraft, pilot recognition time of degraded performance might be quicker and training could be focused on timely and appropriate reactions when faced with unexpected events.

- Performance-based navigation (PBN) could be incorporated to a greater extent. PBN is an initiative of the ICAO (International Civil Aviation Organization, 2013) in which space-based and ground-based equipment with a very high level of precision provides guidance to onboard automation equipment to direct the flight. In these scenarios, demonstrated ability has been shown to keep aircraft within as little as 0.1 mile from a planned path horizontally with a vertical precision that allows input all the way to touchdown on the ground. Applying this technology when faced with other types of malfunctions that require crew attention might reduce the workload of the pilot during critical phases of flight.

Conclusion

This study filled a gap in aviation safety with regard to the impact of incorporating automation into the flight deck. Prior research had looked at individual initiatives in isolation and did not consider the broader impact of technology, human factors, and governance together. This research looked at the systemic impact of automation technology and evaluated current policy guidance to identify areas of opportunity to improve safety. A comprehensive analysis of the system from certification of equipment to implementation by a flight crew revealed numerous areas for improvement. Although it is acknowledged that air travel is already a safe activity (Olson, 2000), the spectacular nature of major accidents and the economic impact of mishaps warrants increased efforts to improve safety. This study provides a context and new perspective on ways to do so.

The study uncovered some results of interest. First the impact of CRM training was expected, but the opportunity for further improvement was not. Similarly, the length of time that human factors research has been performed in the aviation arena made the void in incorporating the results of that research into the technology somewhat unexpected. Additionally, the impact of government and airline oversight was more than anticipated. The Swiss Cheese model provided an excellent vehicle for demonstrating that weakness in policy at the government or operator level often manifested itself in pilots onboard an aircraft trying to react to a situation they felt inadequately prepared for.

The sequential mixed methods approach allowed large numbers of mishaps to be examined and the results of that analysis to provide the focus for a more detailed look at the significance of the findings. The collective findings show multiple areas for research that may improve the safety of the NAS. As the FAA progresses on their plan for a NextGen overhaul of the airspace, the decreasing spacing between aircraft could introduce new risk. This risk could be offset by using the results of this study to direct safety improvements.

9 786121 494133